Random Number Generator on Computers

Random Number Generator on Computers

Naoya Nakazawa
Hiroshi Nakazawa

JENNY STANFORD
PUBLISHING

Published by

Jenny Stanford Publishing Pte. Ltd.
101 Thomson Road
#06-01, United Square
Singapore 307591

Email: editorial@jennystanford.com
Web: www.jennystanford.com

British Library Cataloguing-in-Publication Data
A catalogue record for this book is available from the British Library.

Random Number Generator on Computers

ISBN 978-981-4968-49-2 (Hardcover)
ISBN 978-1-003-41060-7 (eBook)

Contents

Preface

Professor Kiyosi Ito, the founder of stochastic differential equations and Ito calculuses, was the head of the Institute of Mathematical Sciences of Kyoto University in his late years. He was ardent in tutoring physicists and students around on notions related to probability theory and stochastic differential equations. Hiroshi Nakazawa (HN) quotes his words at some occasion,

> *. . . Mathematics is able to discuss any equations. But the deepest and the most fertile equations arise from Physics and Technology. So, we are interested in Physics, . . .*

As a matter of some 40 years ago, memories of HN is not clear about whether Professor Ito just meant to encourage Physicists on his stochastic differential equations, or whether he uttered deep words of wisdom recalling the history of mathematical sciences. We all know Sir Isaac Newton invented differentiation, opened our understanding on the motion of physical objects up to stars, and paved the way for the Industrial Revolution in the European world.

Random numbers have been with us from ancient times, say as coins were tossed or dice were thrown. Yet the modern random numbers arose with the advent of computers, just 80 years ago. Mathematicians, of course, have given long efforts on concepts of random phenomena, but we might well think that the modern swift use of random numbers began with computers. In around 1990, as HN faced the need to convey random number theories at a National College of Technology of Japan to 20-year-old students, he of course referred to the textbook of Professor D. E. Knuth, which stated that random numbers on computers should satisfy some stringent requirements for simulations (see Chapter 1).

These requirements restricted *random numbers on computers* to form (essentially) integral sequences with only finite lengths. The restriction is not suited to the highly developed theories of probability measures that are needed to discuss probabilistic events formed by stochastic processes with continuous sample paths, as arose with Master Kolmogoroff with his cylindrical sets, or as Master Wiener introduced processes now bearing his name.

Random numbers on computers seem to live in a world of integers; Professor Knuth thus begins his monograph *The Art of Computer Programming* with such descriptions. And this is, in fact, an accurate embodiment of the prophet of Professor Ito. This monograph will convince you how such apparent inconveniences of the restriction to integers can show their own exquisite structures of conveniences, with the magical power of Sunzi's theorem.

The investigations in this monograph were not at all foreseen. Some years around 1990, HN faced the need to teach 20-year-old students at a Japanese College of Technology. Without any ideas for other choices, HN jumped to the extensive textbook of Professor Donald E. Knuth. As a newcomer to the subject, HN selected topics that looked digestible to students (and of course to HN himself) and made up a textbook treating the following topics. Please allow HN to quote authors without titles.

(1) Congruential arithmetic of integers
(2) Sunzi's theorem and Euler's function
(3) Elementary group theory and Lagrange theorem
(4) Cyclic groups, subgroups and reduced residue class groups
(5) The condition for cyclic groups
(6) Existence of primitive roots for prime moduluses
(7) Multiplicative congruential (MC) generators modulo 2^r
(8) Spectral tests as introduced by Fishman and Moore
(9) Dieter's algorithm
(10) Finite fields and primitive polynomials
(11) Existence of finite fields
(12) Bose-Chauduri-Hockengem codes
(13) The period of linear congruential equations
(14) The M-sequences on Z_p and Tezuka-Fushimi theorem
(15) Additive Ward generators in the modulus p^r

(16) Additive Brent generators in the modulus 2^r
(17) *Mersenne Twister* of Matsumoto
...

Readers may well think that these headlines are recklessly chosen. Yet, somehow subjects looked interesting to the tutor, and students seemed to enjoy elementary explanations and light exercises showing some fair acceptance. The texbook in Japanese is archived on the homepage of Hirakata Ransu Factory (HRF):

http://www10.plala.or.jp/h-nkzw/

A critic remarked that the textbook treats *too broad and irrelevant classes of subjects.* HN himself shares the same opinion. The truth was that the author could not decipher parts which are essential and which are not. In this chaotic state, HN could do nothing but gather whatever subjects that drew his interest at that time. This state of matter continued until HN's retirement from the college, without any conviction for the future.

After retirement, HN obtained three pieces of great luck. The first was a skill in computer programming. Let $\{x_0.x_1.x_2, \ldots\}$ be a sequence of uniform random numbers. A well-known technique is to plot points $\{P_i = (x_i, x_{i+1})| \ i = 0, 1, 2, \ldots\}$ on a square for $i = 0, 1, 2, \ldots$. This seemingly trivial technique gives versatile geometrical or visual diagnostic judgments on the sequence.

The second was that Naoya Nakazawa (NN), HN's son, joined the research. He was then a graduate student majoring in applied number theory. He obtained a DSc degree and found interest in random number problems on which HN has just found the *Approximation Theorem* to the effect that any random number generators may be approximated by MC generators. Both NN and HN were puzzled by the fact that any two good-looking MC generators could not be combined by Sunzi's theorem to a good-looking generator. As Sunzi's theorem states that a product modulus generator gives a linear combination, or the shuffling, of component modulus random number sequences, this is not reasonable. The puzzle drove HN to examine the criterion of spectral tests anew. He found that spectral test criterions differ from those of regular simplex ones and the latter is a more natural form for the uniformity and independence of random numbers. To re-state, the conclusion was that criterions for MC generators should have the regular simplex forms. The final confirmation was established by NN, who showed that there exist at least two sets of regular simplex criterion passers, #001 and #003, formed by two-component moduluses which are combined by respective good passers of regular simplex criterions. Imagine years of hard efforts without prospects of success. Although there arose further complications that necessitated us to introduce the shortest and the longest edge tests anew on lattices associated with MC random number outputs, they were realized by rather facile and happy works.

There were many ups and downs in the course of these expeditions. We are greatly pleased by our success. With the unsuccessful results of the 20th century, we could try a different approach. Without their extensive efforts, we would have been unable to reach success. Random number problems form difficult subjects, conceptually and mathematically (as represented by Sunzi's theorem). To Professor Knuth, Professor Fishman, Professor Moore, Professor Dieter and all other researchers, NN and HN would like to pay their deepest tribute as predecessors.

We acknowledge specifically decisive contributions of NN on long and bone-breaking investigations for excellent MC generators. Special comments should be made on his contribution to the understanding of the numerical meaning of Sunzi's theorem. His results will soon be disclosed in the form of patents on HRF's home page. We believe that MC random number problems successfully came to an end, except for the endless works to find excellent MC random number generators out of the poker-faced integers. Problems with Sunzi's theorem are now comprehended. This will be the final realization of the prophets of Professor Ito and actual expectation of Professor Knuth.

Please enjoy the problems now settled in this monograph. We are pleased, but looking back at this long-term wander over 30 years, there is little to be proud of. HN would like to express his deepest gratitude to the late Professor Kiyosi Ito, the late Professor Takeyuki Hida, Professor Shinzo Watanabe and, last but not least, Professor John R. Klauder.

Chapter 1

Basic Concepts and Tools

1.1 Random Numbers on Computers as a Sample Process

Courses of University Mathematics for probability start from concepts of probability measures. Discussions of random processes should start with collections of random functions, typically continuous sample functions as the Grand Master Kolmogoroff laid the basis by discretizing sample functions at countable number of time points, and introducing profound ideas of the measure theory. Present gigantic and fertile fields emerged therefrom as the theory of stochastic processes.

We restrict ourselves to problems of random numbers on computers. We are happy to see many simplifications. Numbers on computers are essentially integers in various sense. We thus need only integer sequences placed on discrete time points; the setting gives random numbers as the outcome of a huge dice thrown in computers at discrete times. Yet, we have still to discuss that the dice is fair and the throwing is not deceitful. We present here what we have found within this restricted circumstance. You will be surprised to see that Numbers, which existed from the

Random Number Generator on Computers
Naoya Nakazawa and Hiroshi Nakazawa

ISBN 978-981-4968-49-2 (Hardcover), 978-1-003-41060-7 (eBook)
www.jennystanford.com

beginning of this universe, seem to have prepared neat answers to the present computer problems. But let us not be too hasty, and rewind ourselves from the usual picture that random numbers are reals.

The key looks to be the geometry, or our ability in visual perceptions. Consider a uniformly bound, length T sequence of (real) numbers

$$\{u_1, u_2, u_3, \cdots, u_T \,|\, 0 \leq u_k \leq 1, \quad 1 \leq k \leq T\}$$

as a sequence of uniform random numbers generated on computers. We might reasonably assert:

Proposition 1. (Uniform and independent random numbers) The statistical hypothesis, that a sequence of real random numbers $\{u_1, u_2, \cdots, u_T\}$, with period T and in the unit interval $0 < u_j < 1$, consists of members arising independently of the preceding, is denied most weakly if consecutive 2-tuples of random numbers

$$\{P_j := (u_j, u_{j+1}) |\, 1 \leq j \leq T, \quad P_T := (u_T, u_{T+1}) = (u_T, u_1)\},$$

in the square C in the plane (starting from the origin along axes) with sides of length 1, are distributed uniformly without preference of directions.

(End of Proposition 1)

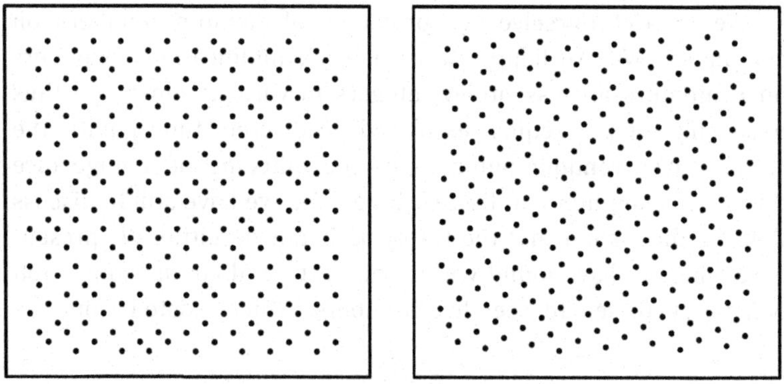

Figure 1

Examples will give us quick convictions. **Figure 1** shows plots of points formed by consecutive 2-tuples from a random number sequence over a period of $T = 250$ as described in **Proposition 1** and from the related multiplicative congruential (MC) (p, z) random number sequence of the same period with the *prime modulus* $p = 251$ and its *primitive root multiplier* $z = 34$. Frames of plots are taken slightly larger than C so as for points at boundaries not to be hidden. Please see **Section 1.3** for the noted correspondence of figures, showing points in this case within the uniform approximation of the precision $1/z = 1/(34)$. Distributions suggest some uniformity and independence of consecutive random numbers. As to your probable feeling that the right MC plot looks too regular, please be patient at this place.

Brief inferences noted above aim to emphasize that we have natural visual ability to discern correlations out of geometrical patterns, just as octopuses, fishes or dinosaurs. The random number technology arose around 1950 with the advent of computers and with the rise of modern computer simulations. The heart of simulations is to find patterns out from the apparently uncorrelated random starts. In this sense, the best random numbers should introduce no correlation by themselves. An obvious ideal is already grasped as uniform and *independent* random numbers. Efforts for their realization on computers revealed gradually that this objective forms one of the hardest problems in modern technology. Happy to say, a breakthrough came around 2008 by the clear recognition of the present authors, that problems allow for powerful and simple approximations in terms of excellent MC random number sequences.

These skipping arguments will be insufficient to persuade you to believe that plots in **Figure 1** is good evidence of uniformity and independence of random numbers, or that plots can be useful at all. The present-day random number generators realize periods $T \approx 2^{52}$. The generation of this whole period on a standard computer will take nearly 2 years, and the plot will give too dense points to be discerned by our eyes. Yet we prove that there are other geometrical methods to comprehend problems. We present ways to realize referred methods. And we shall show their successful and surprising results. The names of operations or weapons are *reformulated spectral tests* and *new edge tests* on *regular simplex criterions*. Our visual ability aided by these new weapons will be seen to have tremendous powers to the end, including excellent system of random numbers distributed on 2- or more dimensional lattices.

1.2 Integer Arithmetic with a Modulus *d*

The sailing to aimed islands cannot be a short voyage. It will require building a new ship itself and its laborious operations may necessitate us to the endurant, perseverant rowing in more than a year on windless sea. But we shall find unexpected beauty of details, ideas and novel flowers. Let us go out to our adventurous and fruitful expeditions.

We proceed expecting readers' standard background in Physics. First we reconfirm our school arithmetic on integers.

Definition 2. (Primes, prime factors, coprime integers) We use x, y, z and in particular d, n for integers.[1]

(A) If $x > 1$ is divided only by 1 and by x itself, x is defined to be a prime.
(B) If an integer $x \geq 2$ is divided by a prime p, then p is said to be a prime factor of x.
(C) The factorization of an integer $x \geq 2$ into prime factors is unique.
(D) If x, y have no common prime factors, they are defined to be coprime.

(End of Definition 2)

All these are well-understood, and might be felt needing no further explanations. In fact, however, we need to make our knowledge firm on the *uniqueness of the decomposition of numbers into prime factors* noted in **(C)** above. To the end we start supplying some miscellaneous terminologies, definitions and confirmations.

[1] *Integers* include 0, and positive as well as negative constituents. *Numbers* are for positive integers $\{1, 2, 3, \cdots\}$.

(a) Let $p > 1$ be a prime. We call p^j for an exponent $j \geq 1$ as a prime factor. Divisors of p^j may be tabulated as $\{1, p^1, p^2, \cdots, p^j\}$, totaling to $j + 1$ in number.
(b) For distinct primes $p \geq 2$ and $q \geq 2$ with respective exponents $j, k \geq 1$, prime factors p^j and q^k are coprime with the greatest common divisor GCD $(p^j, q^k) = 1$.
(c) For distinct primes $\{p_1, p_2, \cdots, p_k\}$ and respective exponents $\{j_1, j_2, \cdots, j_k\}$ which are 1 or larger, the integer $x = p_1{}^{j_1} p_2{}^{j_2} \cdots p_k{}^{j_k}$ is said to be a decomposition of x into prime factors.

We base all notions, *coprime, GCD* or *least common multiple* LCM, on the decomposition into prime factors. These are in fact justified by the uniqueness of prime factorization of integers, and the rigorous base for this operation is only due to the following beautiful corollary of Gauss on the Euclidean algorithm. Please see **Section 3.4** for this beautiful *proof.*

Corollary (EG). Let an integer $x \geq 2$ have *a* decomposition into coprime factors $x = ab$. Then, any prime p dividing x divides either of a or b.

(End of Corollary (EG))

The remaining arguments for the proof run as follows.

Unique Decomposition into Prime Factors Let an integer $x \geq 2$ have a decomposition into prime factors

$$x = p_1{}^{n_1} p_2{}^{n_2} \ldots p_k{}^{n_k},$$

with exponents $\{n_1, n_2, \cdots, n_k\}$ all of which are not less than 1. This factorization is unique.

(Proof) By the assumed order of magnitudes of primes we first consider

$$x = Q_1 Q_2, \quad Q_1 = p_1{}^{n_1}, \quad Q_2 := p_2{}^{n_2} \ldots p_k{}^{n_k}.$$

Factors Q_1, Q_2 are coprime. Thus **Corollary (EG)** ensures that the smallest prime q_1 dividing x divides Q_1, or that $q_1 = p_1$ should be true. We may then introduce $x' = Q_1{}' Q_2, \quad Q_1{}' := p_1{}^{n_1 - 1}$, dividing a prime p_1 off from the problem. If $n_1 - 1 = 0$, we of course take

$x' = Q_2$. This reduction continues obviously, and we reproduce the same

$$x = p_1^{n_1} p_2^{n_2} \cdots p_k^{n_k} = q_1^{n_1} q_2^{n_2} \cdots q_k^{n_k},$$

or the uniqueness of the decomposition into prime factors. ■

Definition 4. (Equivalence relation modulo an integer $d > 0$) Read $x \equiv y \pmod d$ for integers x, y as "x is equivalent to y modulo d," defining that an integer k exists and gives $y = x + kd$.
(End of Definition 4)

Corollary 5. (Equivalence relations modulo d) Let x, y, z, d be integers with the modulus $d > 0$. The equivalence relations modulo d has properties,

(1) reflexive, $x \equiv x \pmod d$,
(2) symmetric, $x \equiv y \pmod d$ implies $y \equiv x \pmod d$,
(3) transitive, $x \equiv y \pmod d$ and $y \equiv z \pmod d$ imply $x \equiv z \pmod d$.

Properties **(A)-(C)** of equivalence relations hold true.

(A) $x \equiv y \pmod d$ holds if and only if x and y have the same remainder when divided by d.
(B) If $xy \equiv xz \pmod d$ holds with x coprime to d, then $y \equiv z \pmod d$ is true. In words, if an integer x is coprime to the modulus d, the equivalence relation including x on both sides may be taken off of x by division.[2]
(C) If $x, y, z, \cdots$ are respectively equivalent to $x', y', z', \cdots$ modulo d, and if $f(x, y, z, \cdots)$ is a polynomial given by additions, subtractions and multiplications of its arguments with integral coefficients, then replacements of any of $x, y, z, \cdots$ with their modulo d equivalents leave $f(x, y, z, \cdots)$ to be equivalent modulo d:

$$\begin{aligned} f(x, y, z, \cdots) &\equiv f(x', y, z, \cdots) \equiv \cdots \\ &\equiv f(x', y', z', \cdots) \pmod d. \end{aligned}$$

[2]We may helpfully think that an integer coprime to d is similar to a non-zero integer in the usual arithmetic. If both sides of a (modular) equation contain the same non-zero factor x, the modular equation may be deleted off of x by division.

Stated verbally: Any polynomial $f(x, y, z, \cdots)$ with integer coefficients remains in the original *equivalence class* when any of its arguments are replaced by modulo-d equivalents.

(Proof) Proofs of (1)-(3) will be manifest.

(A) Divisions of x and y by d give $x = qd + r$ and $y = q'd + r'$ with quotients q, q' and remainders r, r', $0 \le r, r' \le d - 1$. If x and y have the same remainder, we have

$$0 \equiv x - y = d(q - q') + (r - r') \equiv r - r' \pmod{d}$$

with $-d < r - r' < d$, and conclude $r - r' = 0$, or x and y have one and the same remainder if $x \equiv y \pmod{d}$ is true. Conversely, if remainders of x and y divided by d are the same, the equation of division manifestly gives $x = y + kd$ with some integer k, and we have $x \equiv y \pmod{d}$.

(B) If the integer x is coprime to d, then the relation $xy \equiv xz \pmod{d}$ implies $0 \equiv x(y - z) \pmod{d}$. Since x does not contain any prime factors of d, $y - z$ should contain (all prime factors of) d, or $y \equiv z \pmod{d}$. Thus x coprime to d may be divided off from the equivalence $xy \equiv xz \pmod{d}$.

(C) By assumption there are integers $i,\ j,\ k, \cdots$ giving *equations*

$$x' = x + id, \quad y' = y + jd, \quad z' = z + kd, \cdots .$$

Therefore, we typically have

$$x'y = (x + id)y = xy + iyd = xy + \text{(an integral multiple of } d\text{)},$$

because iy is an integer. The same reasoning applies to any $f(x, y,\ z, \cdots)$ in so far as f is formed by addition, subtraction or multiplication only with integer coefficients, and with any number of times of replacements.

(End of Corollary 5)

We emphasize:

(i) Any integers have integral differences, so that the modulus $d = 1$ makes any integers equivalent to each other.

(ii) Note that **(C)** enables us to replace any integer variables of $f(x, y, z, \cdots)$ with their equivalents, typically with their remainders in divisions by d. The manipulation may be performed for any number of times and at any stage of computing. We may of course avoid such replacements, in order to keep arithmetic structures visible, or (in particular on computers) may perform all reduction of magnitude whenever possible to keep integers within a range preventing the overflow of registers.

The use of notions of equivalence classes is thus not at all transcendental, but very practical means. Our most appropriate attitude will be to regard all integers equivalent modulo d to form one and the same entity, to be called *an equivalence class modulo d*, and regard all manipulations cleverly as those performed on all of equivalence class members.

1.3 Arithmetic Structures of T-Periodic Integer Sequences

We now turn to a magic of integers, which is seen by evidence found in our subject, random number sequences on computers. We first describe the plot of the magic. We have seen that random numbers on computers should be regarded as a sequence of length T integers with a sufficiently large T that enables large-scale simulations of today. We restrict ourselves to uniform random numbers (hopefully with independence among consecutive outputs), for other statistics may in principle be obtained by established mathematical transformations. We first introduce the notation of a periodic infinite integer sequence with the period T which are uniformly bound by an integer $z \geq 2$ from above:

$$S(z, T) := \{x_k | \, 0 \leq x_k < z, \quad x_{k+T} = x_k, \quad k = 1, 2, \cdots\}.$$

We assume that integers in $S(z, T)$ are not all zero in order to exclude the trivial case. We then define an infinite sequence of T-periodic uniform rational random numbers:

$$U(z, T) := \{u_k := x_k/z = u_{k+T} \,|\, 0 \leq u_k < 1, \quad k = 1, 2, \cdots\}$$

Define finally a rational number $R(z, T)$ with the expression of a recurring decimal number to the base z with the period T,

$$\begin{aligned}R(z, T) &:= (0.x_1x_2\cdots x_T\; x_1x_2\cdots x_T\; x_1x_2\cdots x_T\;\cdots)_z\\&= x_1z^{-1} + x_2z^{-2} + \cdots + x_Tz^{-T} +\\&\qquad x_1z^{-T-1} + x_2z^{-T-2} + \cdots + x_Tz^{-2T} +\\&\qquad\qquad x_1z^{-2T-1} + x_2z^{-2T-2} + \cdots + x_Tz^{-3T} + \cdots,\end{aligned}$$

which implies that $R(z, T)$ is in the range $0 < R(z, T) \leq 1$.[3]

The rational number $R(z, T)$, expressed by a T-periodic decimal sequence to the base z, admits a representation by an irreducible fraction of integers. The well-known trick gives

$$z^T R(z, T) = x_1z^{T-1} + x_2z^{T-2} + \cdots + x_T + R(z, T),$$

$$R(z, T) = \frac{x_1z^{T-1} + x_2z^{T-2} + \cdots + x_T}{z^T - 1} =: n/d.$$

Here the fraction n/d is assumed irreducible, so that d and n are coprime. Furthermore, d is a factor of $z^T - 1$, so that d and z are also coprime.

The final magic is to show that the set of integers (d, z, n) reproduces, by a recursive system of equations modulo d to the base z for the division n/d, the original integers in $S(z, T)$. The relations in the decimal (or better z-mal) expression are:

$$\begin{aligned}n/d = R(z, T) &= x_1z^{-1} + x_2z^{-2} + \cdots + x_Tz^{-T} +\\&\qquad x_1z^{-T-1} + x_2z^{-T-2} + \cdots + x_Tz^{-2T} +\\&\qquad x_1z^{-2T-1} + x_2z^{-2T-2} + \cdots + x_Tz^{-3T} + \cdots.\end{aligned}$$

We express the fraction n/d, for coprime integers n and d with $1 \leq n < d$, as equations representing procedures of divisions. To the end we introduce first an additional definition $r_0 :\equiv n \pmod{d}$. With quotients $\{x_k\}$ and remainders $\{r_k\}$ in *respective k-th division of the integer n by the integer d for* $k = 1, 2, \cdots$, equations admit

[3] Note that x_k is in the range $0 \leq x_k \leq \bar{z} := z - 1$. Since x_k are assumed to be not all zero, $R(z, T) > 0$ is true. By the summation formula of geometric series, we have $R(z, T)$ is 1 if $x_k = \bar{z}$ for all $k = 1, 2, \cdots$. Thus we have $0 < R(z, T) \leq 1$.

systematic representation:

$$
\begin{aligned}
(z^{-1})\ zr_0 &= x_1 d + r_1,\\
(z^{-2})\ zr_1 &= x_2 d + r_2,\\
(z^{-3})\ zr_2 &= x_3 d + r_3,\\
\cdots\ &\ \cdots\\
(z^{-k})\ zr_{k-1} &= x_k d + r_k,\\
\cdots\ &\ \cdots.
\end{aligned}
$$

These bring out the following clues.

Theorem 6. (A recurrent decimal (or z-mal) sequence and related sequences of integers)[4] Let $S(z, T) := \{x_k |\ 1 \le k \le T,\ \ 0 \le x_k < z\}$ be any length T sequence of zero or positive integers, with the all zero case excluded. Along with $S(z, T)$ define the rational number $R(z, T)$ satisfying[5] $0 < R(z, T) \le 1$ in a decimal (or rather a z-mal) form of period T with the base $z \ge 2$,

$$
\begin{aligned}
R(z, T) :&= (0.x_1x_2x_3 \cdots x_T\ x_1x_2x_3 \cdots x_T\ \cdots)_z\\
:&= x_1z^{-1} + x_2z^{-2} + \cdots + x_Tz^{-T}\\
&\quad + x_1z^{-T-1} + x_2z^{-T-2} + \cdots + x_Tz^{-2T}\\
&\quad\quad + x_1z^{-2T-1} + x_2z^{-2T-2} + \cdots + x_Tz^{-3T} + \cdots.
\end{aligned}
$$

(A) There exist coprime integers d and n satisfying $d \ge 2$ and $1 \le n < d$, wherein d is also coprime to z, that represent $R(z, T)$ as an irreducible fraction $R(z, T) = n/d$.

(B) The original period T sequence

$$U(z, T) := \{u_k := x_k/z |\ k = 1, 2, \cdots, T\}$$

of uniform random numbers is accompanied by the following multiplicative congruential (MC) (d, z) random number

[4] H. Nakazawa and N. Nakazawa: *Designs of uniform and independent random numbers with long period and high precision*, filename *3978erv.pdf* in *www10.plala.or.jp/h-nkzw/indexarchive20jan6.html.*

[5] Define $\bar{z} := z - 1$. The geometric series $\sum_{k=0}^{\infty}(\bar{z}/z)^k = 1$ is readily seen.

sequence $V(d, z, n)$, defined with additional definitions $r_0 :\equiv n \pmod{d}$ and $v_0 := r_0/d$,

$$V(d, z, n) := \{v_k := r_k/d \mid r_k :\equiv nz^k \pmod{d},$$
$$k = 0, 1, 2, \cdots, T - 1\},$$

as the re-construction $\{u_k = v_{k-1} - v_k/z\}$,[6] or as the uniform approximation giving

$$\{0 < v_{k-1} - u_k \leq 1/z \mid k = 1, 2, \cdots, T\}.$$

(Proof)

(A) The representation of $R(z, T)$ by an irreducible fraction n/d, the fact that n/d is irreducible and the fact that d is coprime to z, were all explained in full.

(B) We considered equations of the division n/d as a series in powers of z^{-1}, and compared them with those of $R(z, T)$. Putting $r_0 := n$ as noted, we have the system of equations of division n/d to the base z, as our school mathematics teaches, with the quotient x_k for $k = 1, 2, \cdots$ and the remainder r_k for $k = 0, 1, 2, \cdots$ as tabulated above. Divide all these equations by zd and introduce $\{v_k := r_k/d \mid k = 0, 1, 2, \cdots\}$ which are later identified with the cyclic sequence of multiplicative congruential (MC) random numbers. Then we return back to $v_{k-1} = u_k + v_k/dz$, $k = 1, 2, \cdots$, the starting *uniform random numbers* $\{u_k = x_k/z \mid k = 1, 2, \cdots\} = U(z, t)$. By $0 < v_k \leq 1$ this proves the uniform approximation.[7]

(End of Theorem 6)

The above Theorem is the highway for us to proceed. Since the multiplier z will need to become as large as $z \approx 2^{50}$ or so in simulations nowadays, we may use the MC sequence

$$V(d, z, n) := \{nz^j \pmod{d} \mid j = 0, 1, 2, \cdots\}$$

---

[6]Please note here that the MC random numbers $V(d, z, n)$ have natural staggered counters as $\{v_0, v_1, \cdots, v_{T-1}\}$ while the original random numbers $U(z, T)$ has its natural counters $\{u_1, u_2, \cdots, u_T\}$. Also, please note that in present-day simulations z should be as large as $z \approx 2^{50}$; cf. **Chapter 7**.

[7]There is nothing mysterious about these arguments; they state only the plain fact in division of an integer by another: *A small remainder r_{k-1} is followed by a small next quotient x_k.*

exclusively without hesitation, as approximating the original random number sequence $U(d, z, n)$ to within the precision of 2^{-50}. This choice further allows us great facilities in ways to select tests for excellent (d, z, n) MC generators. Please see **Chapter 7**. We may well be encouraged here that we shall have excellent MC random numbers $V(d, z, n)$ with the true real∗8 precision of $1/z \approx 2^{-50}$, and with the long period $T \approx 2^{52}$.

Chapter 2

Group Structures

The notion of groups was created by Galois. We are not ambitious as to go into the general theory. We shall learn only the most elementary portion of the subject, and that only skippingly. Yet we shall be surprised by the extreme clarity and powerfulness of obtained perspectives.

In arguments of **Theorem 6** we saw that any finite sequence of positive rational numbers smaller than or equal to 1 has the correspondence, in a clear approximation to within the uniform precision $1/z$, to the MC uniform (d, z, n) random numbers. We may recapitulate that *the MC* (d, z, n) *random number sequence may approximate* **any** *uniform random number sequence.* On this perspective we are all justified to concentrate to find a good design for (d, z) generator and to test their precision in the statistical properties of independence. Let us enjoy our expeditions on this extraordinary yacht looking for flowers flourishing even on the most elementary portion of the group theory.

Random Number Generator on Computers
Naoya Nakazawa and Hiroshi Nakazawa

ISBN 978-981-4968-49-2 (Hardcover), 978-1-003-41060-7 (eBook)
www.jennystanford.com

2.1 Reduced Residue Class Groups

Theorem 7. (Reduced residue class groups) For any integer $d \geq 2$ the aggregate of modulo d equivalence classes of integers coprime to d is denoted as $Z_d{}^*$ and called *the reduced residue class group modulo* d. The following axioms hold true for integers in $Z_d{}^*$.

(Axiom 1) Any integers x, y, z in $Z_d{}^*$ are defined of modulo-d multiplication $*$ as well as of the equivalence modulo d, and exhibit properties **(1A)** to **(1C)** noted below.

(1A) $Z_d{}^*$ is closed by the $*$ multiplication: $x * y \in Z_d{}^*$.
(1B) There holds the associative law: $(x * y) * z \equiv x * (y * z)$.
(1C) There holds the commutative law: $x * y \equiv y * x \in Z_d{}^*$.

(Axiom 2) $Z_d{}^*$ has the unit $e \equiv 1 \pmod{d}$ that gives $x{*}e \equiv e{*}x \equiv x$ for any $x \in Z_d{}^*$.

(Axiom 3) Any $x \in Z_d{}^*$ has its inverse $x^{-1} \in Z_d{}^*$ giving $x * x^{-1} \equiv x^{-1} * x \equiv e \pmod{d}$. The aggregate $Z_d{}^*$ will be called *the reduced residue class group modulo d.*

(Proof) (Axiom 1A) Integers $x, y \in Z_d{}^*$ have no common prime factors with d, so that xy as well as $x * y \equiv xy \pmod{d}$ have no common prime factors with d. This proves $x * y \in Z_d{}^*$.

(Axiom 1B) Associative law is true with the usual product $(xy)z = x(yz)$. Taking modulo-d anywhere and any number of times, we have

$$(x * y) * z \equiv x * (y * z).$$

(Axiom 1C) Commutative law is true with the usual product $xy = yx$, and it holds true in modulo d arithmetic.

(Axiom 2) The unit $e \equiv 1 \pmod{d}$ exists as $e = 1 + kd$ with any integer k. We obviously have

$$ex = (1 + kd)x = x + kxd \equiv x \pmod{d},$$

$$xe = x(1 + kd) = x + kxd \equiv x \pmod{d}.$$

(Axiom 3) Take any $x \in Z_d{}^*$. We arrange all representatives of equivalence classes of $Z_d{}^*$ in a line and denote them as $\{a, b, \cdots, g\}$. Multiplication with x gives $\{xa, xb, \cdots, xg\}$ which have no common prime factors with d. To **Corollary 5 (B)** its contraposition proves that $a \not\equiv b$ (mod d) and any $x \in Z_d^*$ imply $xa \not\equiv xb$ (mod d). Therefore, the sequence $\{xa, xb, \cdots\}$ is a permutation of Z_d^*, and there is an element f giving $xf \equiv e$ (mod d). This f is the inverse x^{-1} of $x \in Z_d^*$. ■

In **Corollary 5 (B)** we noted that, if integers a, b, c coprime to d satisfy $ab \equiv ac$ (mod d), we have $b \equiv c$ (mod d). Recognizing that Z_d^* is a group, we may now argue

$$a^{-1} * (a * b) = (a^{-1} * a) * b = e * b = b = a^{-1} * (a * c)$$

$$= (a^{-1} * a) * c = e * c = c$$

by simple arithmetic to obtain the same result.

2.2 Orders, Subgroups and the Theorem of Lagrange

We call the number of elements of a finite group G as the *order* of the group, and denote it sometimes as $\#G$ or $O(G)$. Together with subgroups and the theorem of Lagrange in the title, orders are star players in our arguments, followed by notions of cyclic subgroups and cyclic groups.

We first consider the following on a (not necessarily finite) group G with the $*$ multiplication.

Corollary 8. (Conditions for a subgroup) A subset H of a group G, which is by itself a group w.r.t. the group multiplication $*$ of G, is called a subgroup of G. Necessary and sufficient set of conditions for H to be a subgroup of G is formed by the following:

(a) H is a subset of G.
(b) Any elements $x, y \in H$ give $x^{-1}y \in H$.

(Proof) (Necessity) If H is a subgroup of G, **(a)** and **(b)** certainly hold true. They form a set of necessary conditions.

(Sufficiency) Suppose that **(a)** and **(b)** are true. Taking $x = y$ in **(b)**, we have $x^{-1}y = y^{-1}y = e \in H$, or **(Axiom 2)** for H to be a group. We may thus take $y = e$ in **(b)**, and also that $x \in H$ gives

$x^{-1}e = x^{-1} \in H$, the **(Axiom 3)** holds true. So any $x, y \in H$ give $(x^{-1})^{-1}y = xy \in H$, and H is closed in the multiplication by $*$; **(Axiom A)** is true. Associative and commutative laws are respected by $*$ in G, and in the subset H. ■

We come now within a range for the Theorem of Lagrange.

Theorem 9 (Theorem of Lagrange) If a finite group G has a subgroup H, then the order $g := \#G$ of G is an integral multiple of the order $h := \#H$.

(Proof) We arrange elements of G in a line and denote them as $\{a, b, \cdots, f\}$, and also arrange elements of H as $\{i, j, \cdots, m\}$. Define the **coset** of H as

$$aH := \{ai, aj, \cdots, am\}, \quad a \text{ is any element of } G.$$

We prove here an auxiliary Corollary.

Corollary 10. Let G be a group and H be a subgroup of G. Let any elements a, b of G give cosets aH and bH with the subgroup H. Then aH and bH are either identical as sets, or disjoint with no common elements. (Proof) There are only 2 cases that aH and bH have common element(s) or none. In the former case let $ai = bj$ be the common element with $i, j \in H$. This implies $b = aij^{-1}$ and $bH = a(ij^{-1})H$. Both of i, j are in the subgroup H and gives $k := ij^{-1} \in H$. As kH is just a re-alignment of H, jH is the same set as H. Thus, bH and aH are the same coset. The other alternative is that aH and bH are disjoint with no common element. ■

(Proof of Theorem 9 continued) We take an arbitrary $a \in G$ and construct the coset aH consisting of $h = \#H$ elements. If aH is the whole of G, then $h = \#H = g = \#G$ and **Lagrange Theorem** holds true. If $h = \#H < g = \#G$ is the case, there remains $b \in G$ with $b \notin aH$. The coset bH is disjoint with aH. If aH and bH exhaust G, then $\#G = 2h$ and the assertion is true. In this way we continue to construct disjoint cosets, and the finite $g = \#G$ will eventually be exhausted as $g = \#G = kh = k\#H$ with an integer $k \geq 1$. ■

2.3 Cyclic Subgroups and Cyclic Groups

Hereafter we resume the notation of the product ab for integers in place of $a*b$, and the unit element e will be denoted 1. We continue to take any finite group of order $\#G = g < \infty$, without restriction to the reduced residue class group $G = Z_d{}^*$ alone. Let $a \in G$ be an arbitrary element of G. The sequence of powers of a, $H_a := \{1, a, a^2, \cdots\}$, moves in G and necessarily recurs. If the recurrence occurs in the form $a^i = a^j$ at $0 \le i < j \le g$, we have $a^{j-i} = 1$ for some $0 < j - i \le g$ by the multiplication with $(a^{-1})^i$. Thus the sequence of powers of a has the form

$$1,\ a,\ a^2,\ \cdots,\ a^{h-1},\ a^h = 1,\ a,\ a^2,\ \cdots,\ a^{h-1},\ a^h = 1,\ \cdots$$

and will be called the **Cyclic Sequence** generated by a. We put a definition:

Definition 11. (Order of Group Elements) Any element a of any finite group G has an **order** $O(a) = h$ which is the smallest integer $h \ge 1$ that gives $a^h = 1$. **(End of Definition 11)**

With the order $h = O(a)$ we may also denote

$$H_a := \{1,\ a^1,\ a^2,\ \cdots,\ a^{h-1},\ a^h = 1,\ a^{h+1} = a, \cdots\},$$

if we admit duplicated notations. It is obvious that an element $x = a^j$ and $y = a^k \in H_a$ satisfy the power law $xy = a^j a^k = a^{j+k}$ and has the inverse $y^{-1} = a^{h-k} =: a^{-k}$ by defining negative exponents as usual.

Theorem 12. (Cyclic subgroup) Let $a \in G$ be an arbitrary element of a finite group G, with its order $O(a) = h$. Then h divides the order $g = \sharp G$, and H_a is a subgroup of order h which is named **cyclic subgroup** generated by a.

(Proof) Any elements $x, y \in H_a$ has the form $x = a^j, y = a^k$ and the power law proves $xy^{-1} = a^j a^{h-k} = a^{h+j-k} \in H_a$. Thus H_a is a subgroup of G. The theorem of Lagrange proves that $h = \#H_a$ divides $g = \#G$. ■

Definition 13. (Cyclic group) If an element a of a finite group G has the order $O(a) = g$, then the group is called **Cyclic Group** and the element a is called the generator of the cyclic group G.

(End of Definition 13)

Not all of reduced residue class groups are cyclic. A prime modulus $d = p$ gives the cyclic group $Z_p{}^*$ and its primitive root multipliers construct the whole of $Z_p{}^*$ by their cyclic sequences. It is our keen wish to find a pair of a prime modulus $d = p$ and its primitive root z that give excellent MC random numbers. But tests to find an excellent pair (p, z) demand computing time that increases sharply with the increase of the modulus $d = p$. At some magnitude, say around $d \approx 2^{38}$, non-prime moduluses outweigh in computability. This necessity to change tactics is a significant problem to be tackled in realistic technologies of random numbers. Please see **Chapter 7** for disclosures of some successful examples.

Chapter 3

Designs of MC Generators

3.1 Periods of MC Generators with Prime Moduluses

The first problem of design for the MC integer sequence $\{r_0,\ r_1,\ r_2,\ \cdots\}$ is its length, the period T of repetition. This sequence is in the reduced residue class group $Z_d{}^*$ and is the n-coset of the cyclic subgroup H_z,

$$nH_z :\equiv \{nz^k \pmod{d} | k = 0, 1, 2, \cdots, T-1\}.$$

The period T is the order of z.

Our aim here is to find the necessary smallest portion of the information to be utilized in design, not to go into all details. Even in this modesty, there are many facts to be noted at the start to reduce laborious computing of tests.

Preceding sections stipulate that we need to discuss only MC generators. We should further note restrictions established by Fishman and Moore (1986),[9] that we can only employ *exhaustive (spectral) tests* with no other possible means. Consequently the

[9] G. S. Fishman and L. R. Moore: *An exhaustive analysis of multiplicative congruential random number generators with modulus* $2^{31}-1$, *SIAM Journal on Scientific and Statistical Computing* **7** (1986), pp. 24–45.

Random Number Generator on Computers
Naoya Nakazawa and Hiroshi Nakazawa

ISBN 978-981-4968-49-2 (Hardcover), 978-1-003-41060-7 (eBook)
www.jennystanford.com

largest freedom left for us is to choose a range of moduluses wishing for our luck, and patiently prepare for very long time (which can be years) for computers to sweep over all of candidate primes and multipliers.

Yet, there is some helpful knowledge about the form of the modulus d:

(A) d formed by odd primes;

(A1) a single odd prime $d = p$;

(A2) a product of 2 odd primes $d = p_1 p_2$;

(B) a power of 2, $d = 2^k$.

In **(A2)** we may conceive of any number of primes. Experiences, however, indicate that too small moduluses have little chance to give satisfactory results. We are thus restricted, at least for now, within cases of 1 or 2 odd primes. To our pleasure both of these cases give highly excellent generators; please see **Chapter 7** coming. As to **(B)** we meet the difficulty that the exponent k should be very large now, typically $d = 2^k > 2^{56}$ or larger.[10]

We discuss here on an odd prime modulus $d = p$, with $O(Z_p{}^*) = p - 1$ for the order. In this case the multiplier z has a natural choice for the primitive root. The matter may be rephrased that a prime

---

[10]Many experiences are to be noted here. The composite modulus tactics aim to reduce the computing time of tests on the vague yet plausible expectation that 2 excellent subgenerators (p_1, z_1) and (p_2, z_2) would have larger chances to be combined into an excellent $(d = p_1 p_2, z)$ MC generator, so that the preceding swift tests of (p_1, z_1) and (p_2, z_2) will give more efficient sieves. The trick cannot work with $d = 2^k$. We should also be careful that the composite modulus formed by an odd prime factor and a power of 2, such as $d = 10^k$, should be avoided. This knowledge was obtained around 2008 in Hiroshi Nakazawa and Naoya Nakazawa: *Designs of uniform and independent random numbers with long period and high precision*, March 2-July 8, 2008, with the filename *3978erv.pdf*. Please find this reference in the link given in *http://www10.plala.or.jp/h-nkzw/index.html*, and see its Figures 14–17 for the phenomena and for the reason. This fact implies that the modulus $d = 2^k$ should be used only stand-alone as in Fishman (1990), G. S. Fishman: *Multiplicative congruential random number generators with modulus 2^β: An exhaustive analysis for $\beta = 32$ and a partial analysis for $\beta = 48$*, Mathematics of Computation **54** (1990) pp. 331–344. Here, the time for test computing cannot be reduced by smaller submoduluses just as a single prime modulus. However, we expect that $d = 2^k$ and single prime moduluses may give excellent and efficient MC generators, if our computers admit fast, integer*16 computing. So, the advent of computers with extremely fast integer*16 arithmetic may well remove some of these spells.

p gives a cyclic group $Z_p{}^*$ and any odd prime p has its primitive roots. Since this general proof is not easy, we skip over arguments and present later more visible proofs for special primes of our concern. For general proofs please consult textbooks on Theory of Numbers.[11]

Given a prime p and its primitive root z, our first task is to make notions about cyclic sequences sharper from the point of view of MC (d, z, n) generators. We first note: What are needed for MC random numbers, particularly for *independent* random numbers, are *not the whole* of cyclic or coset sequences from the generator (d, z, n),

$$\{nz^j \pmod{p} | \; j = 1, 2, \cdots\}.$$

We start with the simplest prime-primitive root MC generators.

Lemma 14. (Base sequences and their reverses) We discuss with an odd prime p and its primitive root z fixed, denoting the order of z as $\mathrm{O}(z) = p - 1$. Define the associated *n-primitive root* $\zeta :\equiv -z \pmod{p}$, and construct following 4 finite sequences with the identical length $(p-3)/2$.

(1) The base sequence B on the MC (d, z) generator of a primitive root z:

$$B := \{z^1, z^2, z^3, \cdots, z^{(p-5)/2}, z^{(p-3)/2}\} \pmod{p}.$$

(2) The reverse base sequence $\overline{B}$ on (p, z^{-1}),[12]

$$\overline{B} := \{z^{-1}, z^{-2}, z^{-3}, \cdots, z^{-(p-5)/2}, z^{-(p-3)/2}\} \pmod{p}.$$

(3) Negative base (n-base) sequence B_-, the MC $(p, -z) := (p, \zeta)$ cyclic sequence limited to the length $(p-3)/2$,

$$B_- := \{\zeta, \zeta^2, \zeta^3, \cdots, \zeta^{(p-3)/2}\} \pmod{p}.$$

(4) Reverse of the negative base sequence $\overline{B_-}$ on (p, ζ^{-1}),

$$\overline{B_-} := \{\zeta^{-1}, \zeta^{-2}, \zeta^{-3}, \cdots, \zeta^{-(p-3)/2}\} \pmod{p}.$$

Any one of these 4 base sequences is qualified to be used for MC generators emitting random numbers with the disguise of independence.

---

[11] If you read Japanese, you may be referred to Hiroshi Nakazawa, *Mathematics of Uniform Random Numbers* with the filename *rans1203.pdf*, August 1, 2001-January 12, 2018, archived in the URL *WWW10.plala.or.jp/h-nkzw/indexarchive22june13.html*.

[12] For any multiplier z, the inverse multiplier z^{-1} has the same order $\mathrm{O}(z^{-1}) = \mathrm{O}(z)$. Here, z and z^{-1} are primitive roots modulo p, in particular.

(Proof) A prime-primitive root MC generator (p, z, n) gives outputs, the first half of which runs as

$$\{nz^1, nz^2, \cdots, nz^{(p-1)/2} \equiv -n \bmod (p), \cdots\}.$$

Thereafter follows negatives of all of the first half, which should be discarded as not independent. Deleting uninteresting $z^{(p-1)/2} \equiv -1$, we have the length $(p-1)/2 - 1 = (p-3)/2$ sequence B as the whole of usable random numbers. The multiplier z^{-1} is plainly a primitive root, and $\overline{B}$ may be used for random numbers with the different disguise of independence from B. The n-primitive root $-z$ realizes the MC sequence B_{-} which is B distributed on Euclid spaces with odd-numbered axes taken to the converse direction. The circumstance with $\overline{B_{-}}$ will need no further inferences.

(End of Lemma 14)

The above **Lemma 14** tells us that the MC generation of independent random numbers is ruled by the appearance of ± 1 in the cyclic sequence. This is a general and common guiding rule to all MC generators (d, z, n) other than prime-primitive root construction; MC random number generation is not exclusive works of primitive roots alone. Though circumstances might look complex, we see in **Chapter 5** that all base sequences share identical test valuations. By this fact we need to test only one of them. The knowledge greatly simplifies our coming voyage.

We confirm a small yet helpful corollary here.

Corollary 15. Let G be a cyclic group of the order g, and let a be one of its generator.

(A) General element a^j $(1 \le j \le g)$ of G has the order $h = g/(j, g)$ where (j, g) is the greatest common divisor (GCD) of j and g.

(B) A general element $a^j \in G$ has its inverse $(a^j)^{-1} \equiv a^{g-j}$ with the same order.

(Proof)

(A) As a is a generator of the cyclic group G with $O(G) = g$, the equivalence relation $(a^j)^h = a^{jh} \equiv 1$ holds if jh is a multiple of g. Put $\delta = (j, g)$, $g = g'\delta$ and $j = j'\delta$, with coprime g' and j'. The relation $(a^j)^h \equiv 1$ implies that $jh/g = j'\delta h/(g'\delta) = j'h/g'$ is an integer. The smallest of such h is $h = g' = g/\delta = g/(j, g)$.

(B) Multiplication at once proves $a^j a^{g-j} \equiv a^g \equiv 1$. ■

3.2 Sophie Germain Primes and Naoya Nakazawa Primes

We meet here on particularly useful primes for moduluses of MC random number generation, Sophie Germain (SG) primes and Naoya Nakazawa (NN) primes. They have respective neat structures in their primitive roots and n-primitive roots. We have skipped over the general proof that a prime has its primitive root. We are happy to present here visible proofs that SG primes and NN primes certainly have their primitive roots. The proofs will furnish us with clear ways to sweep over their all relevant primitive roots or n-primitive roots for MC generators.

Theorem 16. (Sophie Germain and Naoya Nakazawa Primes)
(A1) A prime p of the form $p = 2q + 1$ with another prime $q \geq 2$ is called a Sophie Germain (SG) prime. The multiplier $z = 2$ is either a primitive root modulo p with the full order $O(z = 2) = 2q$ or an n-primitive root with the half order $O(z = 2) = q$.

(A2) Primitive roots and their negatives (n-primitive roots) exist one-to-one onto for any SG prime p.

(B) A prime p of the form $p = 4r + 1$ with another prime $r \geq 3$ will be called a Naoya Nakazawa (NN) prime.[13] The multipliers $z \equiv \pm 2$ under the NN prime modulus are primitive roots with the full order, $O(z = 2) = 4r$.

[13]This name corresponds to the neat proof of Naoya Nakazawa shown below. We are not specialized in the theory of numbers, and the name can be inappropriate, owing to our lack of knowledge. The authors will be grateful if readers suggest appropriate and cozier naming.

(Proof) (A1) An SG prime $p = 2q+1$ gives $Z_p{}^* = \{1, 2, \cdots, p-1 = 2q\}$ with the order $O(Z_p{}^*) = 2q$. Lagrange theorem proves that the integer $z = 2$ has its order $O(z)$ in one of $\{2, q, 2q\}$. SG primes start with sets of p and q,

$$(p, q) = (5, 2), (7, 3), (11, 5), \cdots .$$

And $z = 2$ gives $z^2 = 4$, implying $O(2) > 2$. Hence $O(2)$ should be q or $2q$. The case $O(2) = 2q$ implies that $z = 2$ is a primitive root. In the other case $O(2) = q$ we define $z' \equiv -2$, and obtain $(z')^q \equiv (-1)^q \cdot 1 \equiv -1$. This proves that z' is a primitive root, or that 2 is a negative of the primitive root z'; we call this circumstance that 2 is an *n-primitive root* under an SG prime p.

(A2) The above arguments prove that a primitive root z modulo an SG prime $p = 2q + 1$ gives an n-primitive root $z' :\equiv -z$ with the order $O(z') = q$. Conversely, an n-primitive root z' with the order q gives a primitive root $z \equiv -z'$; the *one-to-one onto* correspondence is true.

(B) We first discuss $z = 2$. The prime $r = 3$ gives the smallest NN prime $p = 4 \cdot 3 + 1 = 13$. The reduced residue class group $Z_{13}{}^*$ has the order $O(Z_{13}{}^*) = 12$, and its multiplier $z = 2$ has its order in one of $\{2, 3, 4, 6, 12\}$, giving

$$z^2 = 4,\ z^3 = 8,\ z^4 = 16 \equiv 3,\ z^5 \equiv 6,\ z^6 \equiv 12 \equiv -1,\ z^{12} \equiv 1.$$

Hence $z = 2$ is primitive. The next NN prime is $p = 29$ for $r = 7$. We assume $r \geq 7$ and $p \geq 29$ hereafter. Lagrange Theorem gives $z = 2$ can have its order in $\{2,\ 4,\ r,\ 2r,\ 4r\}$. The condition $r \geq 7$ restricts the order in one of $\{r, 2r, 4r\}$. We prove that $z^{2r} \equiv -1 \pmod{p}$ is true for $z = 2$. If shown, it stipulates that $O(z) = 4r$ and $z = 2$ is a primitive root.[14] Consider the product M:

$$\begin{aligned} M :&= (2 \cdot 1)(2 \cdot 2) \cdots (2 \cdot r)\{2 \cdot (r+1)\}\{2 \cdot (r+2)\} \cdots \{2(2r)\} \\ &= 2^{2r}(2r)! \end{aligned}$$

[14]The neat proof below was communicated to Hiroshi Nakazawa on April 17, 2013, by Naoya Nakazawa.

This has also the following expression modulo $p = 4r + 1$:

$$\begin{aligned} M &= 2 \cdot 4 \cdot \cdots (2r) \cdot (2r+2) \cdot (2r+4) \cdots (2r+2r) \\ &= 2 \cdot 4 \cdots (2r) \cdot (p-(2r-1)) \cdot (p-(2r-3)) \cdots (p-1) \\ &\equiv (-1)^r (2r)! = -(2r)! \pmod{p}. \end{aligned}$$

Here we used that r is an odd prime. These give the equivalence relation modulo p,

$$2^{2r}(2r)! \equiv -(2r)! \pmod{p}.$$

The factor $(2r)!$ is coprime to $p = 4r+1$, and may be divided off from both sides. This gives $2^{2r} \equiv -1 \pmod{p}$, or $O(z) = 4r = q - 1$. Next we consider $z \equiv -1$. From the discussed knowledge of z we have

$$(-z)^1 \equiv -z, (-z)^2 \equiv z^2, \cdots .(-z)^{2r} \equiv z^{2r} \equiv -1.$$

Hence $-z$ is a primitive root and is also a prime to the modulus $p = 4r + 1$. ■

3.3 Sweep over Relevant Multipliers under SG and NN Primes

In **Section 3.1** we saw that a prime p and its any primitive root z give 4 kinds of base sequences of a length $(p - 3)/2$. They are the length $(p - 3)/2$ portion of the cyclic sequences built on z, on the reverse z^{-1}, on the n-primitive root $\zeta \equiv -z$ and on its reverse $\zeta^{-1} \equiv (-z)^{-1} \equiv -z^{-1}$. Any of these may be used for *independent* MC random numbers. We have seen that any SG prime has $z = 2$ as the smallest primitive root or an n-primitive root which is not a primitive root by itself. We also have seen that an NN prime has 2 as the smallest primitive root. These are certainly blessings of these primes on us who are required to sweep over all relevant primes and over their relevant multipliers to exclusively test and find excellent MC generators. Sorry to say, their grace is somewhat ironical in the face of gigantic computing time needed for tests, as we shall experience.

In this section the prime p is restricted to a Sophie Germain (SG) prime or a Naoya Nakazawa (NN) prime. **Corollary 15** at once gives us very simple clues for the sweep:

Corollary 17. (Sweep over SG and NN primes) (A) On an SG prime $p = 2q + 1$ with another prime $q \geq 3$ we need to sweep only over multipliers

$$\{2^{2j-1} \pmod{p} | \; 1 \leq j \leq (p-3)/2\}$$
$$= \{2^{2j-1} \pmod{p} | \; 1 \leq j \leq q-1\},$$

and compute valuations of their tests in *regular simplex criterions.* Obtained valuations are the same irrespective of whether $z = 2$ is a primitive root or an n-primitive root which is in fact *not* a primitive root.[15]

(B) On an NN prime $p = 4r + 1$ with another prime $r \geq 3$ we need to sweep only over multipliers

$$\{2^{2j-1} \pmod{p} | \; 1 \leq j \leq (p-3)/2\}$$

$$= \{2^{2j-1} \pmod{p} | \; 1 \leq j \leq 2r-1\},$$

which are all primitive roots of $p = 4r + 1$.

(Proof) (A) For any SG prime $p = 2q + 1$ with another prime $q \geq 2$, its any primitive root z of order $2q$ corresponds *one-to-one onto* to an n-primitive root of order q. Here, the multiplier $z = 2$ is either a primitive root of $p = 2q + 1$ or a negative, viz. n-primitive root of p with the order q. In case that $z = 2$ is a primitive root, the base sequence

$$B = \{2^{2j-1} \pmod{p} | \; 1 \leq j \leq (p-3)/2\}$$

with $(p-3)/2 = (2q+1-3)/2 = q-1$ suffices for all of relevant multipliers. If $\zeta = 2$ is a negative of the primitive root, i.e., $\zeta = 2$ is an n-primitive root of $p = 2q + 1$, then the base sequence

$$B_- = \{2^{2j-1} \pmod{p} | \; 1 \leq j \leq (p-3)/2\}$$

[15]The test valuations in *regular simplex criterion* are identical for all of 4 base sequences. The subjects require big geometrical arguments needing the whole of Chapter 5. Please be patient at this place with only these suggestive announcements.

again with $(p - 3)/2 = (2q + 1 - 3)/2 = q - 1$, are all of relevant multipliers, putting aside other 3 base sequences. Since all base sequences share the identical (simplex) test valuations as we shall see in **Chapter 5**, the tests on all relevant multipliers on an SG prime modulus need only be performed on the basis sequence B or B_- noted above. Luckily, tests in both cases enjoy identical algebraic construction.

(B) For an NN prime $p = 4r + 1$, we know that $z = 2$ is a primitive root. Hence the base sequence B based on the MC generator $(p = 4r + 1, 2)$ is

$$B = \{2^{2j-1}\ (\text{mod } p) |\ 1 \leq j \leq (p - 3)/2\}$$

with $(p - 3)/2 = (4r - 2)/2 = 2r - 1$. Hence the conclusion follows. ■

3.4 Composite Moduluses and Sunzi's Theorem

Prime moduluses are simple in many respects. However, the problem is that random numbers should have double precision now for large-scale simulations. In computing languages, single precision integers have absolute values from 0 to 2^{31}; they are 4 Bytes (or 2^{32} bit) integers, but a bit should be reserved for the sign. And an integer in double precision should expect their absolute values in the range of 2^{64} or 8 Bytes, with some bit reserved for the sign.

Arithmetic of 8-byte integers is considered in the very design of computer languages. Experientially we feel, however, that we should be careful about many restrictions arising in practices. At present we could somehow work out MC generators giving 2^{56} integer outputs accurately, but to the end we need composite modulus generators, which in turn could only be understood on the basis of Sunzi's theorem. We present accounts on structural implications of Sunzi's theorem to help physical readers to have an intuitive picture of this type of MC random number generators.

We start from an innocently looking property of integral linear combinations of integers.

Corollary 18. (Greatest common divisor GCD) Let m and n be positive integers and S be their aggregate linear combinations with integer coefficients,

$$S := \{Mm + Nn|\ M, N \text{ are positive, 0 or negative integers}\}.$$

Assume that a is the smallest positive integer in S. Then S is the set of all integral multiples of a, and a is the greatest common divisor (GCD) of m and n giving

$$S = \{ka|\ k = 0, \pm 1, \pm 2, \cdots\}.$$

(Proof) The set S includes $m = 1 \cdot m + 0 \cdot n$ and $n = 0 \cdot m + 1 \cdot n$. We assume that the smallest positive integer a in S has the form $a = Mm + Nn$. Let an arbitrary integer in S be $x = M'm + N'n$, and divide x by a into an integral quotient q and the non-negative remainder r. The equation of division is $x = qa + r,\ 0 \le r < a$. This implies the following:

$$0 \le r = x - qa = (M' - qM)m + (N' - qN)n \in S, \quad 0 \le r < a.$$

Since a is the smallest positive integer in S, we have $r = 0$. Thus, any integer x in S is an integral multiple of a. This proves that a divides both m and n with $a \le (m, n)$ where (m, n) is the greatest common divisor GCD of m and n.[16] But the GCD (m, n) of m and n divides $a = Mm + Nn$. Hence $a \ge (m, n)$ is also true, and proves $a = (m, n)$. ■

We thus have:

Theorem 19. (Euclid Algorithm) Any positive integers m, n have their greatest common divisor (m, n) in the form $(m, n) = Mm + Nn$ with integers M, N. **(End of Theorem 19)**[17]

We see the power of Euclid algorithm in revisiting **Corollary (EG)** of Gauss, a small but indispensable clue for the proof of unique factorization property of integers:

---

[16]The symbol (m, n) for the GCD is somewhat confusing with inner products and others. But brevity and conveniences will justify its use.

[17]Integers M, N are not unique. There are integers M', N' fulfilling $M'm + N'n = 0$, e.g., $M' = n, N' = -m$, so that we have $(m, n) = (M + M')m + (N + N')n \equiv \cdots$.

Corollary (EG) If an integer $x \geq 2$ has the factorization $x = ab$ by positive coprime integers $a, b \geq 2$ and if a prime $p > 1$ divides x, then p divides at least one of a, b.

(Proof) There is a dichotomy that p divides a or not. The first possibility sustains the assertion. If p does not divide a, $(p, a) = 1$ holds, because p is divided only by p and 1. Then Euclid algorithm ensures that integers P, A exist and give $Pp + Aa = 1$, which is the same as $Ppb + Aab = b$. As the l.h.s. is divisible by p, p divides b in this case. **(End of Proof)**

The next subject is Theorem of Sunzi, written in a textbook in the era of 439–588 of Southern and Northern Dynasties in China. We describe its simplest form as the Theorem with 2 coprime moduluses for our later use.

Theorem 20. (Theorem of Sunzi) Let the integer d be decomposed into coprime positive factors $d = d_1 d_2$.

(A) For any integer x in the reduced residue class group $Z_d{}^* = Z_{d_1 d_2}{}^*$, there corresponds a pair of integers x_1, x_2 by congruence relations

$$x_1 :\equiv x \pmod{d_1}, \quad x_1 \in Z_{d_1}{}^* \text{ and}$$

$$x_2 :\equiv x \pmod{d_2}, \quad x_2 \in Z_{d_2}{}^*,$$

where $x_1 \in Z_{d_1}{}^*$ and $x_2 \in Z_{d_2}{}^*$ are unique in respective moduluses d_1 and d_2.

(B) Conversely, any pair of integers (x_1, x_2) in the product set $Z_{d_1}{}^* \times Z_{d_2}{}^*$ correspond to an integer $x \in Z_{d_1 d_2}{}^*$ by the system of inverse congruence relations for x,

$$x :\equiv x_1 \pmod{d_1}, \quad x_1 \in Z_{d_1}{}^* \text{ and}$$

$$x :\equiv x_2 \pmod{d_2}, \quad x_2 \in Z_{d_2}{}^*,$$

where x is unique modulo $d = d_1 d_2$. Let the correspondence[18] be denoted as $x :\equiv f(x_1, x_2) \pmod{d}$. Also let integers D_1 and D_2 be

---

[18]This correspondence f and its inverse mapping noted in **(A)** form the group isomorphism of $Z_{d_1 d_2}{}^*$ and the *product group* $Z_{d_1}{}^* \times Z_{d_2}{}^*$.

defined by any form of the Euclidean algorithm

$$1 = D_1 d_1 + D_2 d_2.$$

The correspondence $f(x_1, x_2)$ has an explicit expression

$$x \equiv f(x_1, x_2) :\equiv D_2 d_2 x_1 + D_1 d_1 x_2 \pmod{d_1 d_2}. \qquad (*)$$

(Proof) (A) The statement is obvious by the modular arithmetic.

(B) We first prove modulo d uniqueness of x on the assumption that other integer solution y exists fulfilling

$$y :\equiv x_1 \pmod{d_1}, \quad x_1 \in Z_{d_1}{}^* \text{ and}$$

$$y :\equiv x_2 \pmod{d_2}, \quad x_2 \in Z_{d_2}{}^*.$$

Define $\delta := x - y$. There holds

$$\delta \equiv 0 \pmod{d_1}, \quad \delta \equiv 0 \pmod{d_2}.$$

Hence δ is a multiple of coprime d_1 and d_2, or a multiple of $d_1 d_2 = d$ and vanishes modulo d. This is the uniqueness of the solution modulo d. Integers D_1 and D_2 satisfy $D_2 \equiv d_2^{-1} \pmod{d_1}$ as well as $D_1 \equiv d_1^{-1} \pmod{d_2}$. Hence obviously $(*)$ holds true. ∎

A transcendental and penetrating result of Sunzi's theorem is mathematical as given here. This is all significant to MC random numbers.

Theorem 21. (Euler's function) For any integer $n \geq 1$ the number of positive integers smaller than and coprime to n, i.e., the order $O(Z_n{}^*)$ of elements of a reduced residue class group modulo n, is named *Euler's function* and denoted $\varphi(n)$.

(A) If the integer d has a decomposition into coprime factors $d = d_1 d_2$, then Euler's function shows the *multiplicative* property,

$$\varphi(d_1 d_2) = \varphi(d_1)\varphi(d_2).$$

(B) For a power of a prime p^q there holds

$$\varphi(p^q) = p^q - p^{q-1} = p^q \left(1 - \frac{1}{p}\right).$$

(C) If the integer $n \geq 2$ has the decomposition $n = a^r b^s \cdots f^w$ into prime factors, there holds

$$\varphi(n) = (a^r - a^{r-1})(b^s - b^{s-1}) \cdots (f^w - f^{w-1})$$
$$= n\left(1 - \frac{1}{a}\right)\left(1 - \frac{1}{b}\right) \cdots \left(1 - \frac{1}{f}\right).$$

(Proof) (A) This is the case of **Theorem 20 (B)**. The element $x \in Z_{d_1 d_2}{}^*$ and the element (x_1, x_2) in the product set $(x_1, x_2) \in Z_{d_1}{}^* \times Z_{d_2}{}^*$ correspond one-to-one and onto. Therefore, numbers of elements in both sets are the same, proving $\varphi(d_1 d_2) = \varphi(d_1)\varphi(d_2)$.

(B) The number of integers $\{1, 2, \cdots, p^q\}$ is p^q. Integers containing p are $\{p, 2p, \cdots, p^{q-1}p\}$ and total in number to p^{q-1}. Thus

$$\varphi(p^q) = p^q - p^{q-1} = p^q\left(1 - \frac{1}{p}\right)$$

holds true.

(C) If the integer n has the decomposition into prime factors as said, the multiplicative property and **(B)** prove the assertion. ■

A few values of Euler's function $\varphi(n)$ are readily calculated by counting integers coprime to n:

$$\varphi(1) = \sharp\{1\} = 1, \quad \varphi(2) = \sharp\{1\} = 1, \quad \varphi(3) = \sharp\{1, 2\} = 2,$$

$$\varphi(4) = \sharp\{1, 3\} = 2 = 4(1 - 1/2), \quad \varphi(5) = \sharp\{1, 2, 3, 4\} = 4,$$

$$\varphi(6) = \sharp\{1, 5\} = 2 = \varphi(2)\varphi(3), \quad \cdots.$$

The second significant physical perspective given by Sunzi's theorem is the shuffling structure of MC random number sequences when emitted from composite modulus generators.

Theorem 22. (Shuffling by composite modulus MC generators) Let integers $d_1 > 0$ and $d_2 > 0$ be coprime. The random integer sequence emitted from the MC $(d = d_1 d_2, z, n)$ generator is a shuffling of component random integer sequences

$$\{n_1(z_1)^j \pmod{d_1} |\ j = 0, 1, 2, \cdots\}, \quad n_1 \equiv n \pmod{d_1},$$

$$\{n_2(z_2)^j \pmod{d_2} |\ j = 0, 1, 2, \cdots\}, \quad n_2 \equiv n \pmod{d_2},$$

from MC generators (d_1, z_1, n_1) and (d_2, z_2, n_2), respectively.

(Proof) This is at once from **Sunzi's Theorem 20.** ■

Altogether, we cannot help wondering how integers could foresee the rise of computers in the middle of 20th century. More humanistically, we cannot help admiring works of Euclid, Sunzi, Euler, Gauss, Galois and all excellent geometers and mathematicians in our history, for bringing us to the present perspectives on random number problems on computers.

Please enjoy new knowledge in the following.

Exercise 23. Obtain the number $N_{SG}(p)$ of primitive roots of an SG prime $p = 2q + 1$ with another prime q, and the number $N_{NN}(p)$ of an NN prime $p = 4r + 1$ with another prime r.

(Answer) By **Corollary 15** $N_{SG}(p)$ is the number of integers coprime to the order $p - 1 = 2q$ of the group $Z_p{}^*$. We have

$$N_{SG}(p) = \varphi(2q) = \varphi(2)\varphi(q) = \varphi(q) = q - 1 = (p - 3)/2.$$

Likewise we have

$$N_{NN}(p) = \varphi(4r) = \varphi(4)\varphi(r) = 2\varphi(r) = 2(r - 1) = (p - 3)/2.$$

(End of Exercise 23)

Chapter 4

Lattice Structures

We have investigated possible mathematical structures of MC random number generators. We begin here considerations on their tests. The plan is to let the MC (d, z, n) generator emit random numbers, to collect their *consecutive* l-tuples as points distributed in the l-dimensional Euclidean space, and to examine how the formed pattern is close to the ideal. The plot works beautifully as we shall see. We now successfully have a few excellent MC generators to be disclosed in **Chapter 7**.

4.1 Lattices Accompanied by MC Random Numbers

As **Figure 1** of **Section 1.1** shows, our comprehension on the statistical distribution of emitted MC random numbers is sensed most readily by our sight of the geometrical figures of points formed by consecutive l *integer* outputs. Circumstantially we are helped by the fact that cyclic sequences and their cosets admit no duplication within a period. We concentrate hereafter on geometric features shown by the l consecutive outputs from the MC (d, z, n) sequence,

$$\boldsymbol{V}_j :\equiv \{(nz^j, nz^{j+1}, nz^{j+2}, \cdots, nz^{j+l-1}) \pmod{d} \mid \\ j = 0, 1, 2, \cdots, T-1\}.$$

Random Number Generator on Computers
Naoya Nakazawa and Hiroshi Nakazawa

ISBN 978-981-4968-49-2 (Hardcover), 978-1-003-41060-7 (eBook)
www.jennystanford.com

We take $\{\boldsymbol{V}_j\}$ as position row vectors or points in l-dimensional Euclidean space E_l. By modulo d restrictions on coordinates, we may start by taking them in the hypercube C with edges of length d issuing from the origin along Cartesian coordinates, and then generalize circumstances by taking d-periodic extensions of C to the whole lattice residing in E_l. We start from a firm definition of lattices.

Definition 24. (Lattices and basis vectors) Let l linearly independent[19] row vectors $\{\boldsymbol{e}_1'', \boldsymbol{e}_2'', \cdots, \boldsymbol{e}_l''\}$ be given in E_l with the name of *basis vectors*, issuing from the origin O and giving points as position vectors. The whole set of integral linear combinations of these position vectors and the whole configuration of represented points are named an l-dimensional lattice, and denoted as $L_l(\boldsymbol{e}_1'', \boldsymbol{e}_2'', \cdots, \boldsymbol{e}_l'')$. **(End of Definition 24)**

Vectors with integer coordinates forming a periodic array are in the *integer lattice* formed by all points with integer coordinates, or formed by bases of unit vectors along coordinate axes. Yet this lattice is too fine to be useful, as too many lattice points are left unoccupied by random integer points.[20] Spectral and other tests are based on more efficient choice of the lattice. We restart from an adequate redefinition, as we learned from Fishman and Moore, and others.

Theorem 25. (Lattice formed by l MC outputs) Let $l \geq 2$ be any dimension. Define basis row vectors $\{\boldsymbol{e}_1, \boldsymbol{e}_2, \cdots, \boldsymbol{e}_l\}$:

$$\begin{aligned}
\boldsymbol{e}_1 &:= (1, z, z^2, \cdots, z^{l-2}, z^{l-1}),\\
\boldsymbol{e}_2 &:= (0, d, 0, 0, \cdots, 0, 0),\\
\boldsymbol{e}_3 &:= (0, 0, d, 0, \cdots, 0, 0),\\
&\cdots\cdots\cdots\cdots\cdots\\
\boldsymbol{e}_{l-1} &:= (0, 0, 0, 0, \cdots, d, 0),\\
\boldsymbol{e}_l &:= (0, 0, 0, 0, \cdots, 0, d),
\end{aligned}$$

and define their lattice $L_l := L_l(\boldsymbol{e}_1, \boldsymbol{e}_2, \cdots, \boldsymbol{e}_l)$ in the E_l Euclidean space. Define an l-dimensional cube C therein as a hypercube of

[19] Vectors $\{\boldsymbol{e}_1'', \boldsymbol{e}_2'', \cdots, \boldsymbol{e}_l''\}$ are linearly independent if they span non-vanishing hypervolume, or if they give a matrix with non-vanishing determinant.

[20] The total number of integer points in C is d^l, while consecutive l random numbers of MC (d, z, n) generator can be $\varphi(d) = O(Z_d{}^*) < d$ integer points in C.

sides d issuing from the origin and extending along coordinate axes. Let consecutive l integer outputs of an MC (d, z, n) generator form an l-dimensional row vector in C,

$$\{\boldsymbol{V}_j \equiv nz^j \boldsymbol{e}_1 \pmod d) \mid j = 0, 1, 2, \cdots, T-1\}.$$

Finally extend C, together with points in it formed by consecutive l-tuples of outputs from an MC (d, z, n) generator, to the whole space E_l along all coordinate axes by all d translations. All points of consecutive l-tuples will then occupy almost all of lattice points of $L_l(\boldsymbol{e}_1, \boldsymbol{e}_2, \cdots, \boldsymbol{e}_l)$ in E_l, with the rate of occupation $\varphi(d)/d \approx 1$ and admitting no duplicated seat occupations.

(Proof) Modulo-d arithmetic operations are realized by adding d-translation vectors geometrically. For any $j = 0, 1, 2, \cdots, T-1$ the translated vector $\boldsymbol{V}_j$ in C is expressed as follows with suitable integers $\{c_1, c_2, \cdots, c_l\}$:

$$\begin{aligned}\boldsymbol{V}_j &:= (nz^j \pmod d),\ nz^{j+1} \pmod d),\ \cdots,\ nz^{j+l-1} \pmod d))\\ &= nz^j \boldsymbol{e}_1 - d(c_1, c_2, \cdots, c_l)\\ &= nz^j \boldsymbol{e}_1 - c_1 \boldsymbol{e}_1{}' - c_2 \boldsymbol{e}_2 - \cdots - c_l \boldsymbol{e}_l,\end{aligned}$$

where $\{c_1, c_2, \cdots, c_l\}$ should realize modulo-d retractions of $\boldsymbol{V}_j$ to within C, and $\boldsymbol{e}_1{}'$ is the d-translation vector along the 1st axis of coordinates,

$$\boldsymbol{e}_1{}' := (d, 0, 0, \cdots, 0) = d\boldsymbol{e}_1 - z\boldsymbol{e}_2 - z^2\boldsymbol{e}_3 - \cdots - z^{l-1}\boldsymbol{e}_l.$$

The resultant vector $\boldsymbol{V}_j$ takes the form

$$\begin{aligned}\boldsymbol{V}_j = (nz^j - dc_1)\boldsymbol{e}_1 - (c_2 - c_1 z)\boldsymbol{e}_2\\ -(c_3 - c_1 z^2)\boldsymbol{e}_3 - \cdots - (c_l - c_1 z^{l-1})\boldsymbol{e}_l.\end{aligned}$$

This is an integral linear combination of $\{\boldsymbol{e}_1, \boldsymbol{e}_2, \cdots, \boldsymbol{e}_l\}$, and is a lattice point of $L_l(\boldsymbol{e}_1, \boldsymbol{e}_2, \cdots, \boldsymbol{e}_l)$. Consider an arbitrary set of d-translations that may let $\boldsymbol{V}_j$ go out of C, with $c_1'd, c_2'd, \cdots, c_l'd$ translations to 1st, 2nd, $\cdots$, lth coordinates:

$$\begin{aligned}&c_1'\boldsymbol{e}_1{}' + c_2'\boldsymbol{e}_2 + c_3'\boldsymbol{e}_3 + \cdots + c_l'\boldsymbol{e}_l\\ &= c_1'(d\boldsymbol{e}_1 - z\boldsymbol{e}_2 - z^2\boldsymbol{e}_3 - \cdots - z^{l-1}\boldsymbol{e}_l) + c_2'\boldsymbol{e}_2 + c_3'\boldsymbol{e}_3 + \cdots + c_l'\boldsymbol{e}_l\\ &= dc_1'\boldsymbol{e}_1 + (c_2' - c_1'z)\boldsymbol{e}_2 + (c_3' - c_1'z^2)\boldsymbol{e}_3 + \cdots + (c_l' - c_1'z^{l-1})\boldsymbol{e}_l.\end{aligned}$$

Calculations conclude that any vector $\boldsymbol{V}_j{}'$, which may be in or out of C, takes the form

$$\begin{aligned}\boldsymbol{V}_j{}' = &\{nz^j + d(c_1{}' - c_1)\}\boldsymbol{e}_1 + \{(c_2{}' - c_1{}'z) - (c_2 - c_1 z)\}\boldsymbol{e}_2\\ &+\{(c_3{}' - c_1{}'z^2) - (c_3 - c_1 z^2)\}\boldsymbol{e}_3 + \cdots\\ &+\{(c_l{}' - c_1{}'z^{l-1}) - (c_l - c_1 z^{l-1})\}\boldsymbol{e}_l.\end{aligned}$$

Thus, all d-translations of points formed by l-consecutive MC outputs take their seats in the lattice $L_l(\boldsymbol{e}_1, \boldsymbol{e}_2, \cdots, \boldsymbol{e}_l)$, because $c_1{}', c_2{}', \cdots, c_l{}'$ may be assigned of any integer values and $\boldsymbol{e}_2, \boldsymbol{e}_3, \cdots, \boldsymbol{e}_l$ have their integer coordinates, running over all independent values. The coefficient of $\boldsymbol{e}_1$ is restricted, however, to be coprime to d. Thus, seats provided by the lattice $L_l(\boldsymbol{e}_1, \boldsymbol{e}_2, \cdots, \boldsymbol{e}_l)$ are occupied by consecutive l-tuples of MC (d, z, n) random number outputs with the occupation ratio of $\varphi(d)/d < 1$.[21] ■

4.2 Basis Vectors of a Lattice

A lattice admits innumerably many sets of its basis vectors. The set of basis vectors $\{\boldsymbol{e}_1, \boldsymbol{e}_2, \cdots, \boldsymbol{e}_l\}$ given in the preceding **Section 4.1** is nice in showing arithmetic relations to consecutive l random integer outputs from (d, z, n). However, this and many other sets of basis vectors will not fit well to show geometrical features of lattice point arrangements. Let us restart by more general reflections on lattice structures, assuming that a lattice is given just as an array of points in the space with periodic structures.

Take the dimension $l = 2$. If a line in E_2 has 2 lattice points on it, we call it a lattice line. In fact, there are infinitely many lattice points on this lattice line. Take any pair of neighboring lattice points[22] O and A, and define the first basis vector $\boldsymbol{e}_1 := \overrightarrow{\mathrm{OA}}$.[23] Then take the

---

[21]A specific feature of the set of basis vectors introduced above is to be noted. The first axis direction summarizes the l consecutive MC random numbers for its coordinates though, of course, reductions by (mod d) should be followed. This feature of Cartesian coordinates will be seen to play a special role later.

[22]Two lattice points on a lattice line are *neighboring* if there are no other lattice points between them.

[23]Note that in the particular lattice taken in relation for Cartesian coordinates, the vector $\overrightarrow{\mathrm{OA}}$ is in the direction of the first coordinate axis.

2nd lattice line, which is parallel and neighboring to the first.[24] Take any lattice point B on it, and define $\boldsymbol{e}_2 := \overrightarrow{\mathrm{OB}}$. Then 2 vectors $\{\boldsymbol{e}_1, \boldsymbol{e}_2\}$ form a set of basis vectors for the lattice in E_2. Observe that the parallelogram spanned by $\{\boldsymbol{e}_1, \boldsymbol{e}_2\}$ contains no lattice points inside or on its edge lines excepting vertices. If this parallelogram is moved parallel to vectors $\boldsymbol{e}_1$ and/or $\boldsymbol{e}_2$ so as for its vertex O to itinerate all lattice points, the whole Euclidean plane E_2 is tiled with this parallelogram without gap or overlap. This is true with any choice of basis vectors, so that every set of basis vectors span a parallelogram with the same area. In the case of consecutive 2 MC (d, z) outputs, we saw a possible form of basis vectors

$$\boldsymbol{e}_1 = (1, z), \quad \boldsymbol{e}_2 = (0, d).$$

Their obvious determinant is d, so that this identical area is d.

Next we take a lattice L_3 in the 3-dimensional space E_3 for cases with $l = 3$ consecutive outputs. If a 2-dimensional plane has 3 lattice points which are not on a line, then we name it a lattice plane. There are infinitely many lattice points on this plane and they form by themselves a 2-dimensional sublattice. Basis vectors of this 2-dimensional lattice are constructed by taking 3 lattice points O, A and B as before, and forming $\boldsymbol{e}_1 = \overrightarrow{\mathrm{OA}}$ and $\boldsymbol{e}_2 = \overrightarrow{\mathrm{OB}}$. We then take a 2nd lattice plane parallel and neighboring to the 1st, take an arbitrary lattice point C on this 2nd plane, and define $\boldsymbol{e}_3 = \overrightarrow{\mathrm{OC}}$. Then $\{\boldsymbol{e}_1, \boldsymbol{e}_2, \boldsymbol{e}_3\}$ form a set of basis vectors for L_3. In the case of the MC lattice for 3 consecutive outputs, we have seen that they have a possible form

$$\boldsymbol{e}_1 = (1, z, z^2), \quad \boldsymbol{e}_2 = (0, d, 0), \quad \boldsymbol{e}_3 = (0, 0, d)$$

with the obvious determinant $d^2 > 0$. Therefore, any set of basis vectors of this lattice have one and the same volume d^2.

In dimensions $l \geq 4$ our vision cannot easily imagine the geometry of lattices, but noted procedures to construct basis vectors may be extended without difficulties. In the present status of random number technology, we need cases up to $l = 6$. This upper limit for l is set on us by computing-time difficulties. From the general comprehension that MC generators can approximate any

[24]Two parallel lattice lines are neighboring if no lattice points exist between them.

random numbers on computers, we strongly expect that computers with extremely high speed in the integer arithmetic, based possibly on new physical principles, will give solutions for higher dimensions. Anyway, thus obtained forms of basis vectors for MC random number problems span one and the same hypervolume d^{l-1}.

We shall later need transformations of basis vectors in a lattice in general dimension. *Unimodular transformations* introduced below is for this need.

Theorem 26. (Unimodular transformations) Call an $l \times l$ *integer* matrix U_l with $\det U_l = \pm 1$ *unimodular*. Let a lattice L_l be defined with a set of basis vectors $\{\mathbf{e}_1{}', \mathbf{e}_2{}', \cdots, \mathbf{e}_l{}'\}$ as $L_l := L_l(\mathbf{e}_1{}', \mathbf{e}_2{}', \cdots, \mathbf{e}_l{}')$. Then, the set of vectors $\{\mathbf{e}_1{}'', \mathbf{e}_2{}'', \cdots, \mathbf{e}_l{}''\}$ defined by the linear transformation of any unimodular matrix $U_l = (u_{ij})$,

$$\begin{aligned}
\mathbf{e}_1{}'' &= u_{11}\mathbf{e}_1{}' + u_{12}\mathbf{e}_2{}' + \cdots + u_{1l}\mathbf{e}_l{}', \\
\mathbf{e}_2{}'' &= u_{21}\mathbf{e}_1{}' + u_{22}\mathbf{e}_2{}' + \cdots + u_{2l}\mathbf{e}_l{}', \\
&\cdots\cdots\cdots\cdots\cdots\cdots, \\
\mathbf{e}_l{}'' &= u_{l1}\mathbf{e}_1{}' + u_{l2}\mathbf{e}_2{}' + \cdots + u_{ll}\mathbf{e}_l{}',
\end{aligned}$$

is a set of basis vectors of the same lattice,

$$L_l = L_l(\mathbf{e}_1{}', \mathbf{e}_2{}', \cdots, \mathbf{e}_l{}') = L_l(\mathbf{e}_1{}'', \mathbf{e}_2{}'', \cdots, \mathbf{e}_l{}'').$$

Conversely, two arbitrary sets of basis vectors of any lattice are unimodular transformations of each other. With a lattice L_l fixed, there is one-to-one and onto correspondence between all sets of basis vectors of L_l and all of $l \times l$ unimodular matrices.

(Proof) Introduce matrix notations for a set $\{\mathbf{e}_1{}', \mathbf{e}_2{}', \cdots, \mathbf{e}_l{}'\}$ of basis vectors, for any other set $\{\mathbf{e}_1{}'', \mathbf{e}_2{}'', \cdots, \mathbf{e}_l{}''\}$ of basis vectors[25], and likewise for U_l as depicted below explicitly in matrix forms. These give a compact expression $E_l{}'' = U_l E_l{}'$ for noted unimodular transformation:

$$E_l{}' := \begin{pmatrix} \mathbf{e}_1{}' \\ \mathbf{e}_2{}' \\ \vdots \\ \mathbf{e}_l{}' \end{pmatrix}, \quad E_l{}'' := \begin{pmatrix} \mathbf{e}_1{}'' \\ \mathbf{e}_2{}'' \\ \vdots \\ \mathbf{e}_l{}'' \end{pmatrix}, \quad U_l := \begin{pmatrix} u_{11} & u_{12} & \cdots & u_{1l} \\ u_{21} & u_{22} & \cdots & u_{2l} \\ \cdots & \cdots & \cdots & \cdots \\ u_{l1} & u_{l2} & \cdots & u_{ll} \end{pmatrix}.$$

[25]Matrices $E_l{}'$ and $E_l{}''$ introduced below will be called *basis matrices* hereafter.

(1) First, remind ourselves that any sets of basis vectors, or any basis matrices, are obtained by choosing lattice vectors suitably, as noted at the beginning of this section.

(2) We next note that a unimodular matrix U_l is regular by $\det U_l = \pm 1$, has its inverse $U_l{}^{-1}$, and $U_l{}^{-1}$ is also unimodular, because $\det U_l{}^{-1} = 1/(\det U_l) = \pm 1$ holds true and the cofactor formula proves that $U_l{}^{-1}$ is an integer matrix.

(3) We finally note that a basis matrix $E_l{}'$ of a lattice L_l and any unimodular matrix U_l gives $E_l{}'' := U_l E_l{}'$ which is by itself a basis matrix of L_l. This is because row vectors of $E_l{}'$ gives any lattice vector of L_l as their integral linear combinations, while those rows of $E_l{}'$ are integral linear combinations of rows of $E_l{}''$ by $E_l{}' = U_l{}^{-1} E_l{}''$. Hence rows of $E_l{}''$ expresses any lattice vector of L_l, which proves that $E_l{}''$ is a basis matrix of L_l.

(4) Reversible matrix relations $E_l{}'' = U_l E_l{}'$ and $U_l = E_l{}''(E_l{}')^{-1}$ prove the one-to-one onto correspondence at once. ■

4.3 Dual Lattices for Spectral Tests

We use abbreviated notations of $E_l(\boldsymbol{e}_1, \boldsymbol{e}_2, \cdots, \boldsymbol{e}_l)$ such as $E_l(d, z)$ or E_l hereafter without comments, if there will arise no fear of confusion. Consider the lattice of l tuple integer outputs from MC (d, z, n) generator, with basis row vectors $\{\boldsymbol{e}_1, \boldsymbol{e}_2, \cdots, \boldsymbol{e}_l\}$. We define dual basis row vectors $\{\boldsymbol{f}_1, \boldsymbol{f}_2, \cdots, \boldsymbol{f}_l\}$ with the notation of inner product $(\boldsymbol{a}, \boldsymbol{b})$ for vectors $\boldsymbol{a}$ and $\boldsymbol{b}$ as follows.

$$(\boldsymbol{e}_i, \boldsymbol{f}_j) = \delta_{ij}, \quad 1 \le i, j \le l.$$

We denote the transposition of a matrix A as $\overline{A}$, and the unit matrix as I. Then the dual lattice defined above is summarized in a simple form of a matrix equation

$$E := \begin{pmatrix} \boldsymbol{e}_1 \\ \boldsymbol{e}_2 \\ \cdots \\ \boldsymbol{e}_l \end{pmatrix}, \quad F := \begin{pmatrix} \boldsymbol{f}_1 \\ \boldsymbol{f}_2 \\ \cdots \\ \boldsymbol{f}_l \end{pmatrix}, \quad dI = E\overline{F} = F\overline{E},$$

$$F = d\overline{E^{-1}} = d(\overline{E})^{-1}.$$

We call F the dual basis matrix. The relation $\det dI = d^l$ proves that F has its determinant independent of l:

$$|\det(F\overline{E})| = |\det F| \cdot |\det \overline{E}| = d^l, \qquad \det F = d.$$

Corollary 27. (The relation between basis vector matrix E and its dual matrix F) Said correspondence is as follows:

$$\begin{aligned}
\boldsymbol{e}_1 &:= (1, z, z^2, \cdots, z^{l-2}, z^{l-1}) & \boldsymbol{f}_1 &:= (d, 0, 0, 0, \cdots, 0, 0),\\
\boldsymbol{e}_2 &:= (0, d, 0, 0, \cdots, 0, 0) & \boldsymbol{f}_2 &:= (-z, 1, 0, 0, \cdots, 0, 0),\\
\boldsymbol{e}_3 &:= (0, 0, d, 0, \cdots, 0, 0) & \boldsymbol{f}_3 &:= (-z^2, 0, 1, 0. \cdots, 0, 0),\\
&\cdots\cdots\cdots\cdots\cdots & &\cdots\cdots\cdots\cdots\cdots,\\
\boldsymbol{e}_l &:= (0, 0, 0, 0, \cdots, 0, d) & \boldsymbol{f}_l &:= (-z^{l-1}, 0, 0, 0, \cdots, 0, 1).
\end{aligned}$$

(Proof) Inner products will prove the whole assertion at once. ■

For a basis matrix E consider its unimodular transformation $E' = UE$ by a unimodular matrix U. This is $E = U^{-1}E'$, and the dual basis matrix F for E satisfies

$$dI = F\overline{E} = F\overline{U^{-1}E'} = F\,\overline{E'}\,\overline{U^{-1}}.$$

Multiplying these relations with $\overline{U^{-1}}$ from the left and with $\overline{U}$ from the right, we have

$$dI = F'\overline{E'}, \quad F' = \overline{U^{-1}}F.$$

Let $L_l(d, z) := L_l(\boldsymbol{e}_1, \boldsymbol{e}_2, \cdots, \boldsymbol{e}_l)$ denote the lattice in the space E_l spanned by consecutive l-tuples of MC (d, z, n) outputs as denoted in **Corollary 27**. Summarize its basis matrix briefly as $E_l(d, z)$. Its dual basis matrix $F_l(d, z)$ has *integer components* as **Corollary 27** shows. These matrices are related to each other as follows

$$F_l(d, z) = d\,\overline{E_l(d, z)^{-1}} = d\{\overline{E_l(d, z)}\}^{-1}.$$

In the obvious abbreviated form we may write this as $F = d(\overline{E})^{-1}$.

We have now a general perspective on the dual basis matrix $F_l(d, z)$ and related dual lattice.

Theorem 28. (Dual lattices) Let $E = E(d, z)$ be the MC (d, z, n) basis matrix for the lattice $L_l = L_l(d, z)$. Let U be any $l \times l$ unimodular matrix and denote $E' = UE$.

(A) Define dual matrices F and F' on E and E', respectively as $F = d\overline{E^{-1}}$ and $F' = d\overline{E'^{-1}}$. Then F' is a unimodular transformation of F with the unimodular matrix $U^* := \overline{U^{-1}}$, namely $F' = U^*F$ holds true.

(B) A set, formed by a basis matrix E of a lattice L and all of unimodular matrices, is known to give *all* basis matrix of L. Completely in parallel, the set of a basis matrix F and the set of all of unimodular matrices form all of (dual) basis matrices of a lattice L^* which is named the dual lattice of L.

(Proof) (A) The dual basis matrix F' of E' is

$$F' = d\overline{(E')^{-1}} = d\overline{(U\,E)^{-1}} = d\overline{E^{-1}U^{-1}} = d\overline{U^{-1}}\,\overline{E^{-1}} = \overline{U^{-1}}F = U^*F.$$

(B) Statements are evident. ■

The obtained relation denotes a neat correspondence that enables us to explain the technology of *spectral tests* in the next section.

4.4 Spectral Tests of MC (d, z) Lattices

As already noted, visual inspections give by far the richest information on distribution properties of random numbers. MC random numbers are cyclic or coset sequences or, as we came to recognize more accurately, 4 types of base sequences, all with non-duplicating property adapted best for such geometrical inspections. Even with these favorable properties, all of the available tests face difficulties which are summarized as the *shortest vector problems of lattices*. There are problems of huge computing time. We have no other way to evade the difficulty, however, than to carry out laborious *exhaustive tests* as named by Fishman and Moore (1986),[26] but with some helpful experiences and technical know-how.

At this place it will not be useless to give short structural reflections on our difficult expedition to random number generation problems. In the beginning, we had great luck recognizing that MC generators should exclusively be taken as random number generators on computers. This motivated us to learn modular arithmetic,

[26]Please see footnote 9 at p. 19.

and we were led to reduced residue class groups. Lagrange theorem admirably simplified periodic structures. Composite moduluses necessitated us to learn brilliant Sunzi's theorem and Euclid algorithm to help beautiful Gauss proof and clear up mysterious Euler's function. And we were led finally to embed the whole of random number geometries in lattices. We are now within one-step to structures of spectral tests which, even after 37 years from the basis laid by Fishman and Moore, cannot be said that penetrating comprehension is around. Let us start from the simplest dimension $l = 2$ of 2 consecutive outputs from an MC (d, z) generator.

Take the 2-dimensional plane E_2 and consider the 2-dimensional lattice $L_2(d, z)$ spanned by basis vectors $\boldsymbol{e}_1 = \overrightarrow{\mathrm{OA}}$ and $\boldsymbol{e}_2 = \overrightarrow{\mathrm{OB}}$. We consider the triangle spanned by these 2 basis vectors, which has the name *2-simplex spanned by* $\{\boldsymbol{e}_1, \boldsymbol{e}_2\}$. The case of parallelogram of area d spanned by $\{(1, z), (0, d)\}$ proves the area of this simplex is $d/2$.

Consider triangles in **Figure 2**, showing cases spanned by basis vectors $\overrightarrow{\mathrm{OA}} = \boldsymbol{e}_1$ and $\overrightarrow{\mathrm{OB}} = \boldsymbol{e}_2$. Imagine the lattices they form; we need to supplement innumerable number of lattice lines passing through the vertices of these triangles. Consider the lattice line passing through B, which is parallel to the lattice line passing through A. Between these 2 lattice lines no lattice points intervene, and they are neighboring lattice lines. The system $\{\boldsymbol{e}_1, \boldsymbol{e}_2\}$ of basis vectors determines dual basis vectors $\{\boldsymbol{f}_1, \boldsymbol{f}_2\}$. As $\boldsymbol{f}_2$ is orthogonal to $\boldsymbol{e}_1$, the vertex B has the height $\mu = |(\boldsymbol{e}_2, \boldsymbol{f}_2)|/\|\boldsymbol{f}_2\| = d/\|\boldsymbol{f}_2\|$ to base line $\overline{\mathrm{OA}}$, with $\|\boldsymbol{f}_2\|$ for the length of the vector $\boldsymbol{f}_2$.

This comprehension is summarized as follows:

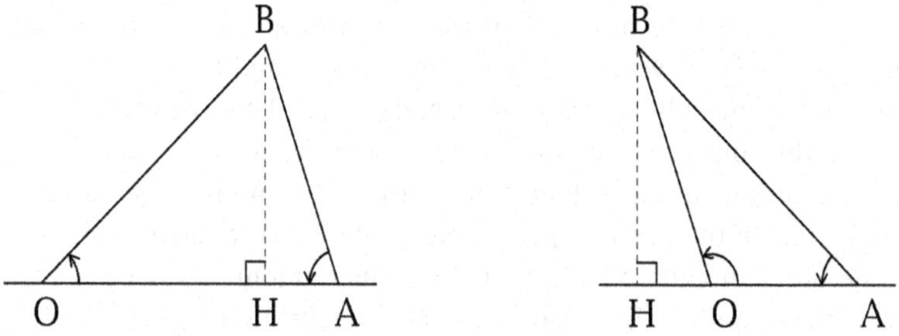

Figure 2 Principles of spectral tests.

Corollary 29. (2nd-degree spectral tests) Consecutive 2-tuples from the MC generator (d, z) give the lattice $L_2(d, z)$ with $\mu_2{}^{\max}(d, z)$ for the largest distance between parallel and neighboring lattice lines in $L_2(d, z)$,

$$\mu_2{}^{\max}(d, z) := d/\|\boldsymbol{f}_{\min}\|.$$

Here $\|\boldsymbol{f}_{\min}\|$ denotes the length of the shortest non-zero lattice vector $\boldsymbol{f}_{\min}$ in the dual MC lattice $L_2{}^*(d, z)$.

(Proof) Let $\{\boldsymbol{e}_1{}', \boldsymbol{e}_2{}'\}$ be any set of basis vectors of $L_2(d, z)$. Any parallel and neighboring lattice lines given by this set of basis vectors were seen to have their distance $\mu = d/\|\boldsymbol{f}_k{}'\|$ with $k = 1, 2$. This relation is true with any dual basis vectors, and the largest distance between parallel neighboring lattice lines given by them. Thus the largest distance is given by the shortest length of basis vectors in $L_2(d, z)^*$. The shortest (dual) basis vector may manifestly be replaced with the shortest (dual) lattice vector. ■.

A neat generalization of this clue to $l \geq 2$ of $L_l(d, z)$ is at once.

Theorem 30. (Principle of degree l spectral tests) Let $L_l(d, z)$ be the lattice in which points of l consecutive MC (d, z) random numbers take their seats. Parallel and neighboring lattice hyperplanes of $(l - 1)$-dimension have their largest distance $\mu_l{}^{\max}(d, z)$ given by

$$\mu_l{}^{\max}(d, z) := d/\|\boldsymbol{f}_{\min}\|$$

where $\boldsymbol{f}_{\min}$ is the lattice vector in the dual lattice $L_l{}^*(d, z)$ with the shortest positive length $\|\boldsymbol{f}_{\min}\|$.

(Proof) Let $\{\boldsymbol{e}_j{}' = \overrightarrow{\mathrm{OA}_j} \mid j = 1, 2, \cdots, l\}$ be any set of basis vectors of the MC lattice $L_l(d, z)$, and $\{\boldsymbol{f}_1{}', \boldsymbol{f}_2{}', \cdots, \boldsymbol{f}_l{}'\}$ be corresponding dual basis vectors of the dual lattice $L_l{}^*(d, z)$. Take any basis vector, which for simplicity is assumed to be $\boldsymbol{e}_1{}' = \overrightarrow{\mathrm{OA}_1}$. The end point A_1 of this vector faces the $(l - 1)$-dimensional hyperplane spanned by remaining $(l - 1)$ basis vectors, namely the point set in E_l given by

$$\{c_2\overrightarrow{\mathrm{OA}_2} + c_3\overrightarrow{\mathrm{OA}_3} + \cdots + c_l\overrightarrow{\mathrm{OA}_l} \mid c_2, c_3, \cdots, c_l \text{ are real numbers}\}.$$

Denote the height of A_1 to this hyperplane as μ. Since the dual basis vector $\boldsymbol{f}_1$ is orthogonal to all vectors spanning this hyperplane, we have

$$\mu = |(\boldsymbol{e}_1, \boldsymbol{f}_1)|/\|\boldsymbol{f}_1\| = d/\|\boldsymbol{f}_1\|.$$

This relation holds true with any basis vector and its dual, together with any other set of basis vectors and their duals. It will be obvious that the shortest basis vector of a (dual) lattice is the shortest (dual) lattice vector. ■

The last clue of spectral tests is the search of the shortest dual lattice vector. Here is a brilliant algorithm stated by Dieter.

Theorem 31. (Dieter) A necessary and sufficient condition, for a dual vector $\boldsymbol{f}$ in the lattice $L_l{}^*(d, z)$ with *Cartesian coordinates* $\boldsymbol{f} = (y_1, y_2, \cdots, y_l)$ is the following:

$$y_1 + zy_2 + z^2y_3 + \cdots + z^{l-1}y_l \equiv 0 \pmod d. \qquad (*)$$

(Proof) We choose dual lattice basis vectors for $L_l{}^*(d, z)$:

$$\boldsymbol{f}_1 := (d, 0, 0, 0, \cdots, 0, 0),$$

$$\boldsymbol{f}_2 := (-z, 1, 0, 0, \cdots, 0, 0),$$

$$\cdots\cdots\cdots\cdots\cdots,$$

$$\boldsymbol{f}_l := (-z^{l-1}, 0, 0, 0, \cdots, 0, 1).$$

If $\boldsymbol{f} = (y_1, y_2, \cdots, y_l)$ is in $L_l{}^*(d, z)$, there are integers $\{c_1, c_2, \cdots, c_l\}$ giving

$$\boldsymbol{f} = c_1\boldsymbol{f}_1 + c_2\boldsymbol{f}_2 + \cdots + c_l\boldsymbol{f}_l$$

$$= (c_1d - c_2z - c_3z^2 - \cdots - c_lz^{l-1}, c_2, c_3, \cdots, c_l).$$

This implies:

$$y_1 + zy_2 + z^2y_3 + \cdots + z^{l-1}y_l = c_1d \equiv 0 \pmod d.$$

Hence $(*)$ is necessary. Conversely, assume that $(*)$ holds true. Then integer Cartesian coordinates $(y_1, y_2, \cdots, y_l)$ of $\boldsymbol{f}$ should have an integer k giving the equation

$$y_1 + zy_2 + z^2y_3 + \cdots + z^{l-1}y_l = kd,$$

and

$$\boldsymbol{f} = (kd - zy_2 - z^2y_3 - \cdots - z^{l-1}y_l, y_2, y_3, \cdots, y_l)$$

$$= k\boldsymbol{f}_1 + y_2\boldsymbol{f}_2 + y_3\boldsymbol{f}_3 + \cdots + y_l\boldsymbol{f}_l,$$

implying that $\boldsymbol{f} \in L_l{}^*(d, z)$ is true. Thus $(*)$ is sufficient. ■

We have arrived at key relations, **Theorems 30** and **31** in the theory of spectral tests of an MC (d, z) generator. To proceed further, we need *regular simplex criterions* which have not been considered in MC random numbers so far. This forms the final key of our new success. Please see **Section 5.3**.

Chapter 5

Regular Simplexes and Regular Lattices

Discussions up to the present leave many other subjects to be desirably discussed. Yet, we would turn here to different aspects to push ourselves into a new important recognition. Let $L_l(d, z)$ be the lattice in which l consecutive MC (d, z) outputs are situated, and denote $\{\overrightarrow{\mathrm{OA}}, \overrightarrow{\mathrm{OB}}, \cdots, \overrightarrow{\mathrm{OF}}\}$ for its l basis vectors. We may say that these vectors give the *l-simplex* of hypervolume $d^{l-1}/l!$ in the linear hull fixed by $l+1$ end points $\{\mathrm{O, A, B}, \cdots, \mathrm{F}\}$. We prefer here the reference to simplex rather than to parallelepiped, for simplexes are geometrically simpler and more tractable in dimensions $l \geq 4$, particularly. And, happy to say, problems of tests of MC random number lattices enable us to concentrate only on aspects of **regular simplexes**. Please see below how the change frees us from images of complicated higher dimensional geometries.

5.1 Construction of Regular Simplexes and Regular Lattices

An l-simplex is a cone with solid inside and outer hyperplanes in the space E_l. Since we are interested in lattices, we shall think more generally that the outer linear frame formed by edges, or more

Random Number Generator on Computers
Naoya Nakazawa and Hiroshi Nakazawa

ISBN 978-981-4968-49-2 (Hardcover), 978-1-003-41060-7 (eBook)
www.jennystanford.com

simply outer $l + 1$ points or vertices $\{O, A, B, \cdots, F\}$, also form an l-simplex together with their linear hull. We think here only on regular l-simplexes. They visibly admit the following inductive construction by compasses for $l = 2, 3, \cdots$, and remind us of a visible symmetry related to the MC generator (d, z).

Definition 32. (Inductive definition of regular simplexes) (A) In the 1-dimensional Euclid space E_1 a line element $\overline{OA}$ with length $a > 0$ is named a *1-simplex*. We assume that 1-simplex is in the first coordinate axis issuing from the origin O.

(B) In the plane E_2 and for any 1-simplex $\overline{OA}$ of length $a > 0$ on the 1st coordinate axis, we draw circles of radius a from O and A. Choosing a point B of intersection, we define the frame of the regular triangle ΔOAB and define its linear hull as the *regular 2-simplex*. We take the 2nd coordinate axis orthogonal to the 1st, choosing a direction from 2 possibilities. It stems from O to the direction perpendicular to the 1st axis. **(C)** In the 3-dimensional Euclidean space E_3 and for any regular 2-simplex ΔOAB laid in E_3 with edges of length $a > 0$, we draw spheres of radius a centered at O, A and B. Choosing one point C of intersections of spheres, we define a *regular 3-simplex* as the linear hull spanned by points O, A, B and C with the geometrical shape of a regular tetrahedron. The 3rd coordinate axis stems from O perpendicularly to the 2-simplex. **(D)** In the space E_l with dimension $l \geq 2$, let an $(l - 1)$-dimensional regular simplex of edge length $a > 0$ with l vertices O, A, B, $\cdots$, E be laid. Drawing hyperspheres of radius a from O, A, B, $\cdots$, E, and taking one point F of intersection, a *regular l-simplex* is defined inductively as the linear hull spanned by $l + 1$ vertices O, A, B, $\cdots$, F. The l-th coordinate axis takes one possibility in 2, and stems from O perpendicularly to the $(l - 1)$-regular simplex. **(E)** If a set of l basis vectors span a regular l-simplex, the lattice formed by their integral linear combinations is defined to be an l-dimensional *regular lattice*.

(End of Definition 32).

A regular 2-dimensional lattice is also called the *triangular lattice*. A regular 3-dimensional lattice has the name *face-centered cubic lattice* in physics. The following trivial Reminder is the basis of *simplex tests* for valuations of MC (d, z) random number generators.

Reminder 33. (Neighbors in regular lattices) In a regular lattice, *any* lattice point and its nearest neighbors have one and the same distance to be denoted as ε^*. Likewise, any lattice point and its next nearest neighbors have one and the same distance to be denoted as ε_l^{**}. There holds $\varepsilon_l^{**} > \varepsilon_l^*$.

(Proof) Let A and B be any lattice points in a regular lattice L_l. Geometrical circumstances around A and around B are identical by their construction by compasses. Hence the statements are obvious. Quantitative values for distances $\varepsilon_l^* < \varepsilon_l^{**}$ will be seen in **Section 5.3** as the ideal case of MC (d, z) lattices. ■

5.2 Actual Lattices Associated with MC Generators

We show the **Figure 3** occupying 3 pages below depicting the distribution of 2 tuples of consecutive MC (p, z) integer outputs for prime-primitive root generators with a variety of 2nd-degree spectral test valuations. Plots are depicted in the square C, the linear extension of which is taken slightly larger than d. As generators, we took small modulus $d = p$ to ensure the distribution to be perceptible to our eyes. For higher dimension $l \geq 3$ depiction is difficult, and we ask readers to use their imagination. If these plots help your *feel* on the relation of spectral test valuations and actualities, we shall be happy.

We refer to some technological details. We intentionally chose small moduluses for the recognizability to our eyes. Plots are depicted in the square C, the linear extension of which is taken slightly larger than d. Plots are associated with their second degree spectral test valuations $\rho = \rho_2(p, z)$. Please construct your *experience* on the relation of geometrical distributions and spectral test valuations. If you let these PDF files very slowly, points are to appear in the order as then are generated, as they are plotted in the order of MC generation.

After Pioneers Fishman and Moore, put the conventional criterion that an MC generator is (2nd-degree) possible if the valuation does not exceed 1.25. Construct your recognition from

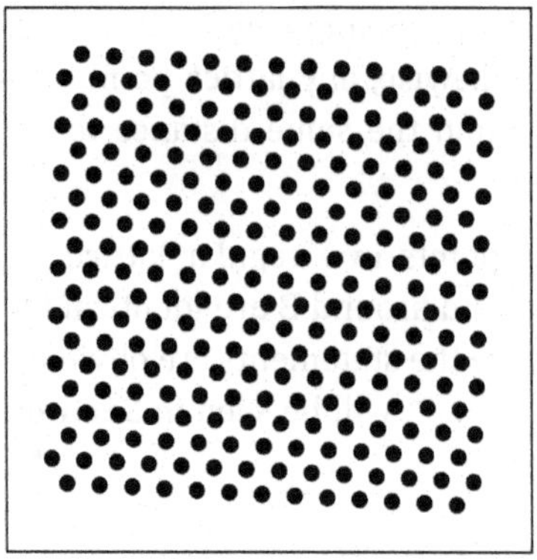

$\rho = 1.0503 \quad (257, 27)$

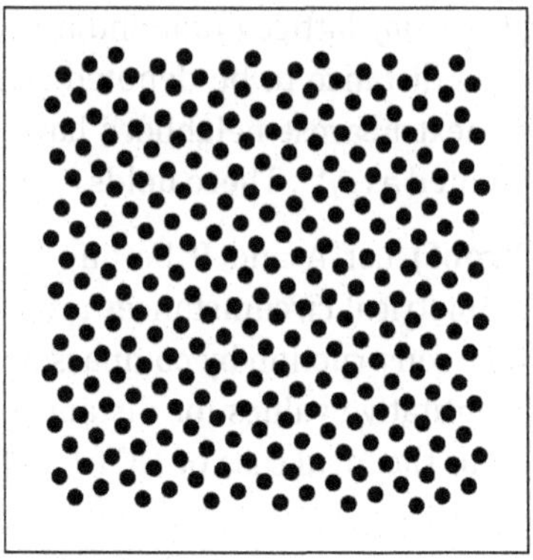

$\rho = 1.0983 \quad (283, 83)$

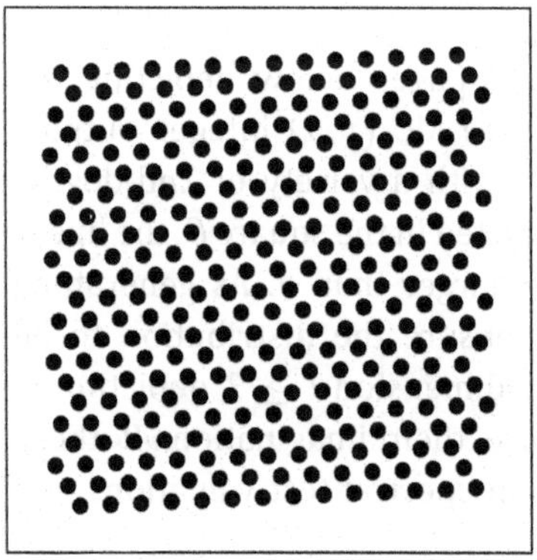

$\rho = 1.1459 \quad (317, 245)$

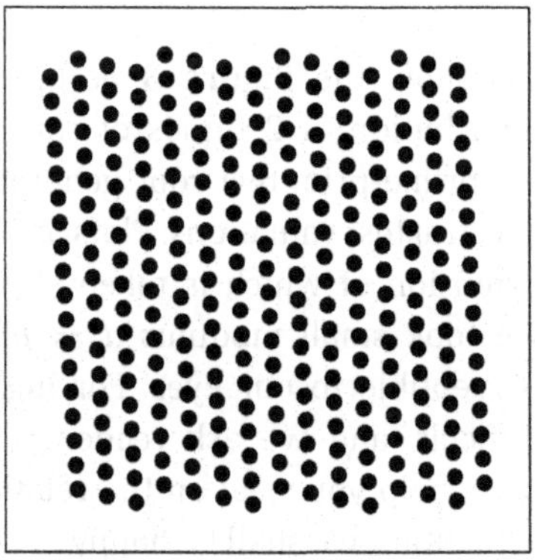

$\rho = 1.1982 \quad (281, 266)$

Figure 3 *(continued)*.

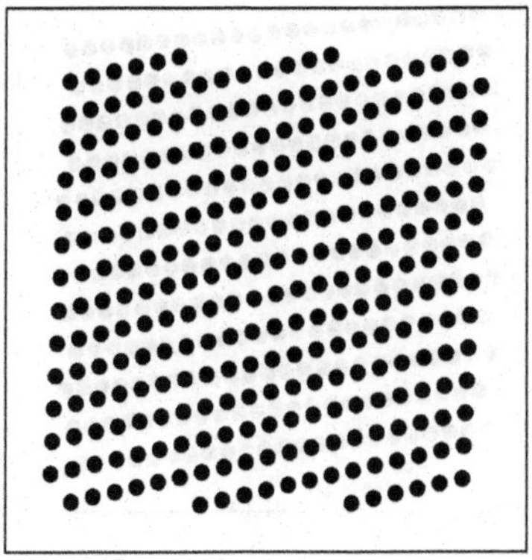

$\rho = 1.2491 \quad (277, 20)$

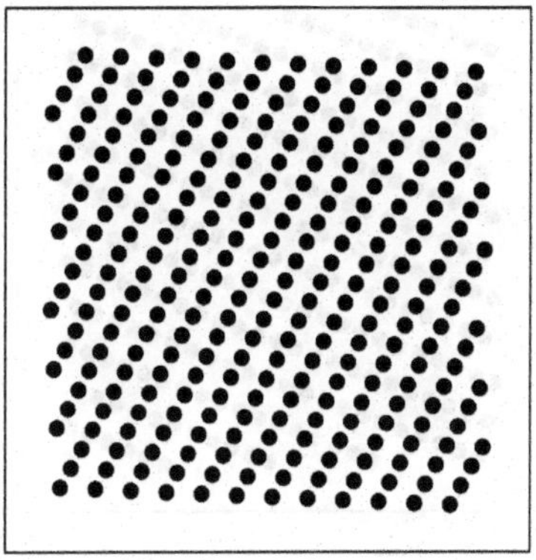

$\rho = 1.3012 \quad (283, 113)$

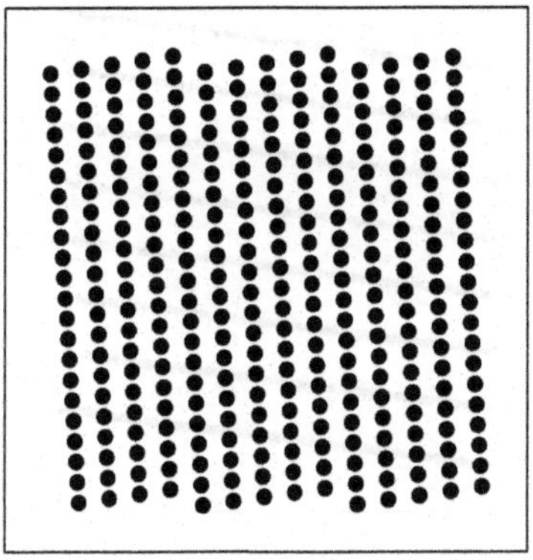

$\rho = 1.3501 \quad (311, 297)$

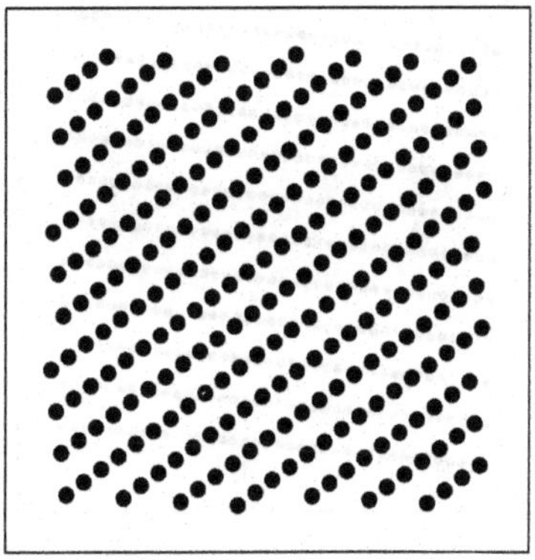

$\rho = 1.3947 \quad (251, 76)$

Figure 3 (*continued*).

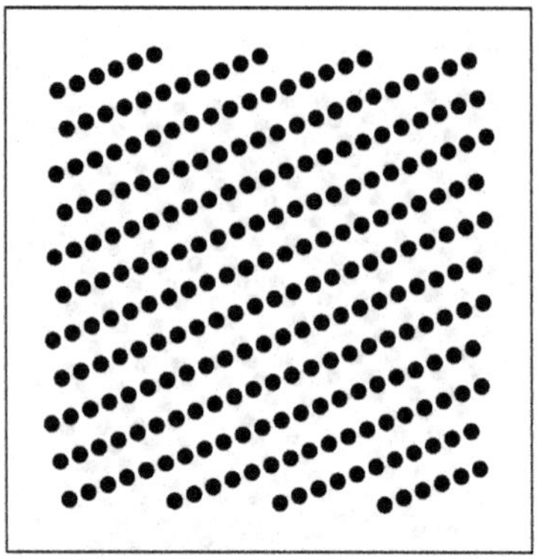

$\rho = 1.4547 \quad (251, 46)$

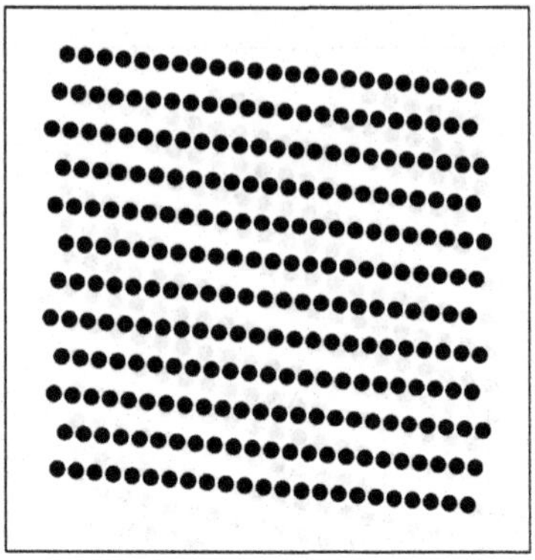

$\rho = 1.4959 \quad (281, 117)$

$\rho = 1.992 \quad (419, 381)$

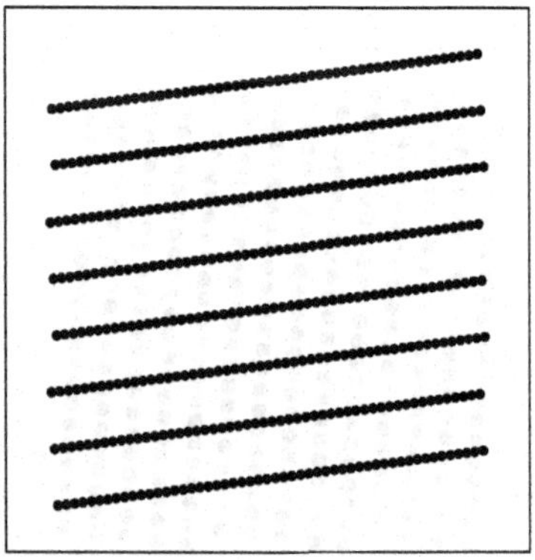

$\rho = 2.7283 \quad (419, 262)$

Figure 3 Points of consecutive 2 outputs.

these plots that this 2nd-degree criterion is, and presumably higher dimensional criterions will be, all reasonable.

We note 2 problems here, however. One is that our visual perception should grasp sufficiently many emitted points in the period. We may see the motion picture of seat-taking by random numbers in **Figure 3** which are plotted in chronological order of generation. If these PDF files are reproduced sufficiently slowly, points in plots will appear precisely in the order of generation of random numbers. One way to realize this might well be to let our computer display these PDF files, while $N \gg 1$ Command Prompts on the same computer are simultaneously performing simple endless computations at the same time. Then **Figure 3** will show the seat-taking processes of their points to appear in chronological order. Trials will convince us that a considerable portion of periods should be depicted for us to conceive of shapes of the lattices in **Figure 3**. But here arise difficulties of computing time. Suppose that we have an MC generator with the period $T \approx 2^{54}$. The generation of such a large number, $2^{54} \times 2/5$ say, of random numbers, and that one by one without skipping, will take years. This implies practical impossibility for visual judgments. The other somewhat minor problem is that we shall need some microscope to magnify and *see* a very small portion of the lattice in C. In short, we need to devise tests which will nicely indicate the geometrical closeness with numeral valuations (and needing only a computable time, of course). The problem is: On what quantities associated with regular lattices should these valuations be taken? Experiences persuade us that there are ways for us to choose, with respective efficiencies and difficulties in computability.

5.3 Regular Simplex Criterions

Let $l \geq 2$ be the dimension and assume the set of l (basis) vectors below to form a regular l-simplex with the hypervolume $d^{l-1}/l!$ in the space E_l.

$$\begin{aligned}
\boldsymbol{e}_1{}^* &:= \overrightarrow{\mathrm{OQ}_1} = (b, a, a, \cdots, a, a),\\
\boldsymbol{e}_2{}^* &:= \overrightarrow{\mathrm{OQ}_2} = (a, b, a, \cdots, a, a),\\
\boldsymbol{e}_3{}^* &:= \overrightarrow{\mathrm{OQ}_3} = (a, a, b, \cdots, a, a),\\
&\cdots\cdots\cdots\\
\boldsymbol{e}_{l-1}{}^* &:= \overrightarrow{\mathrm{OQ}_{l-1}} = (a, a, a, \cdots, b, a),\\
\boldsymbol{e}_l{}^* &:= \overrightarrow{\mathrm{OQ}_l} = (a, a, a, \cdots, a, b).
\end{aligned}$$

Without loss of generality $a > 0$ and $b \neq a$ are assumed.

We easily obtain the determinant formed by these vectors:

$$\begin{vmatrix} b & a & a & \cdots & a & a \\ a & b & a & \cdots & a & a \\ a & a & b & \cdots & a & a \\ \cdot & \cdot & \cdot & \cdot & \cdot & \cdot \\ a & a & a & \cdots & b & a \\ a & a & a & \cdots & a & b \end{vmatrix} = \begin{vmatrix} s & s & s & \cdots & s & s \\ a & b & a & \cdots & a & a \\ a & a & b & \cdots & a & a \\ \cdot & \cdot & \cdot & \cdot & \cdot & \cdot \\ a & a & a & \cdots & b & a \\ a & a & a & \cdots & a & b \end{vmatrix}$$

$$= \begin{vmatrix} s & 0 & 0 & \cdots & 0 & 0 \\ a & c & 0 & \cdots & 0 & 0 \\ a & 0 & c & \cdots & 0 & 0 \\ \cdot & \cdot & \cdot & \cdot & \cdot & \cdot \\ a & 0 & 0 & \cdots & c & 0 \\ a & 0 & 0 & \cdots & 0 & c \end{vmatrix} = sc^{l-1},$$

$$s := (l-1)a + b, \quad c := b - a.$$

Conditions **(A)** and **(B)** noted below ensure that these vectors give a regular l-simplex, hence span a regular lattice $L_l^*(d)$:

(A) The absolute value of the determinant,[27] formed by these vectors as rows, is d^{l-1}.

(B) These vectors, together with vectors formed by all pairs of their end points $\{Q_1, Q_2, \cdots, Q_l\}$, have one and the same length,

$$\|\boldsymbol{e}_i^*\| = \|\boldsymbol{e}_j^* - \boldsymbol{e}_k^*\|, \quad 1 \le i, j, k \le l, \quad j \ne k.$$

Let us think over the circumstance to start with. Above assumed l vectors have one and the same length $\|\boldsymbol{e}_j^*\| = \{(l-1)a^2 + b^2\}^{1/2}$. And for $j \ne k$ inner products are all $(\boldsymbol{e}_j^*, \boldsymbol{e}_k^*) = (l-2)a^2 + 2ab$, so that they form one and the same angle with others. Therefore, the condition, that all ${}_{l+1}C_2$ pairs are equidistant,

$$\|\boldsymbol{e}_k^* - \boldsymbol{e}_j^*\|^2 = \|\overrightarrow{Q_jQ_k}\|^2 = 2(b-a)^2 = \|\boldsymbol{e}_j^*\|^2, \quad j \ne k, \quad 1 \le j, k \le l,$$

is the same as the condition that remaining ${}_lC_2 = l(l-1)/2$ pairs have the identical squared distance $\|\boldsymbol{e}_j^*\|^2$. We assume $b = \xi a$, and

[27]The absolute value $d^{l-1} > 0$ of this determinant proves that these vectors span a hypervolume d^{l-1}, and that they are linearly independent.

consider this condition **(B)**. Since a is not 0, **(B)** gives the equation $(l-1)+\xi^2 = 2(\xi-1)^2$ for ξ, viz.

$$\xi^2 - 4\xi - l + 3 = 0, \quad \xi = \xi_\pm := 2 \pm (l+1)^{1/2}.$$

Any of $\pm$ signs gives the aimed construction of a regular l-simplex. We put $\xi_\pm := 2 \pm (l+1)^{1/2}$. The unknown ξ_+ is simply positive for all $l = 2, 3, \cdots$, and may readily be treated. On the other hand, ξ_- changes its sign by l:

$$(l=2) \quad \xi_- = 2 - 3^{1/2} > 0 \qquad (l=3) \quad \xi_- = 2 - 4^{1/2} = 0$$

$$(l=4) \quad \xi_- = 2 - 5^{1/2} < 0 \qquad (l=5) \quad \xi_- = 2 - 6^{1/2} < 0$$

$$(l=6) \quad \xi_- = 2 - 7^{1/2} < 0$$

This circumstance complicates the treatment. We thus take the simpler case of ξ_+ with positive a and b.[28]

Assume $l \geq 2$, take $\xi = \xi_+ > 3$, and put $\Lambda := (l+1)^{1/2}$. We may then write $\xi = 2 + \Lambda$. Since we have

$$s = (l-1)a + b = (l-1+\xi)a = (l+1+\Lambda)a = a\Lambda(\Lambda+1),$$

$$c = b - a = (\xi - 1)a = a(\Lambda+1),$$

the condition **(A)** implies

$$d^{l-1} = |a\Lambda(\Lambda+1)\cdot\{a(\Lambda+1)\}^{l-1}| = a^l\Lambda(\Lambda+1)^l.$$

This solves $a > 0$ as

$$a = \frac{d^{(l-1)/l}}{\Lambda^{1/l}(\Lambda+1)} = \frac{d^{(l-1)/l}(\Lambda-1)}{\Lambda^{1/l}(\Lambda+1)(\Lambda-1)} = \frac{d^{(l-1)/l}(\Lambda-1)}{\Lambda^{1/l}l}.$$

Similarly, b is:

$$b = \xi a = (\Lambda+2)a = \frac{d^{(l-1)/l}(\Lambda+2)(\Lambda-1)}{\Lambda^{1/l}l} > 3a.$$

We now have all relevant quantities. The largest distance between parallel and neighboring lattice hyper planes was denoted $\mu_l{}^*(d)$. It may be obtained in terms of d and l with the unit vector $\boldsymbol{e}_0 :=$

[28]This is the case that the regular simplex lies in the portion of E_l where all coordinates are positive. We add that $\xi = \xi_-$ is also interesting, in particular in the dimension $l = 3$ vectors then span the tetrahedron with a clear correspondence to the physical way of taking coordinates as the *face-centered cubic lattice*.

$(1, 1, \cdots, 1)/l^{1/2}$ issuing from the point O and the vector $\overrightarrow{OQ_j}$ for any $1 \le j \le l$ as their inner product as follows:

$$\mu_l{}^*(d) = |l^{-1/2}\{(l-1)a + b\}| = l^{-1/2}(l - 1 + \xi)a$$
$$= l^{-1/2}(\Lambda^2 + \Lambda)\frac{d^{(l-1)/l}(\Lambda - 1)}{\Lambda^{1/l}l} = \frac{d^{(l-1)/l}\Lambda(\Lambda+1)(\Lambda-1)}{l^{3/2}\Lambda^{1/l}}$$
$$= \frac{d^{(l-1)/l}\Lambda^{(l-1)/l}}{l^{1/2}} = d^{(l-1)/l}(l+1)^{(l-1)/(2l)}l^{-1/2}.$$

This at once gives the spectral tests in regular simplex criterions. See the summary in **Theorem 34 (A)**.

Let $\varepsilon_l{}^*(d)$ be the edge length of the regular l-simplex of the nearest neighbor distance in the regular l lattice $L_l{}^*$. Obviously it is given by

$$\varepsilon_l{}^*(d) = \|\boldsymbol{e}_j{}^*\| = \{2(b-a)^2\}^{1/2} = 2^{1/2}(b-a).$$

Thus we have

$$\varepsilon_l{}^*(d) = 2^{1/2}\frac{d^{(l-1)/l}\{(\Lambda+2)-1\}(\Lambda-1)}{\Lambda^{1/l}l}$$
$$= 2^{1/2}\frac{d^{(l-1)/l}(\Lambda+1)(\Lambda-1)}{\Lambda^{1/l}l}$$
$$= 2^{1/2}\frac{d^{(l-1)/l}l}{\Lambda^{1/l}l} = 2^{1/2}d^{(l-1)/l}(l+1)^{-1/(2l)}.$$

We denote $\varepsilon_l{}^{**}(d)$ for the next nearest distance between lattice points of the l-dimensional regular lattice $L_l^*(d)$. The construction by a compass gives

$$\varepsilon_l{}^{**}(d) := 2\mu_l{}^*(d).$$

We soon need the ratio $\gamma_l{}^* := \varepsilon_l{}^*(d)/\varepsilon_l{}^{**}(d)$ which is d-independent,

$$\gamma_l{}^* = \frac{\varepsilon_l{}^*(d)}{\varepsilon_l{}^{**}(d)} = \frac{2^{1/2}d^{(l-1)/l}(l+1)^{-1/(2l)}}{2d^{(l-1)/l}(l+1)^{(l-1)/(2l)}l^{-1/2}} = 2^{-1/2}\left(\frac{l}{l+1}\right)^{1/2}.$$

This increases as $l \to \infty$ within the bound $\gamma_l^* < 2^{-1/2} \approx 0.7071$.

All results are summarized in the theorem below.

Theorem 34. (Regular simplex criterions) The l-dimensional *regular lattice* $L_l{}^*(d)$ gives criterions $\{\mu_l{}^*(d) \mid 2 \le l \le 6\}$ noted below in **(A)–(C)**.

(A) Between parallel and neighboring $(l-1)$-dimensional lattice hyperplanes in $L_l{}^*(d)$, the largest distance is

$$\mu_l{}^*(d) = d^{(l-1)/l}(l+1)^{(l-1)/(2l)}l^{-1/2}.$$

Its values for $l = 2, 3, \cdots$ are respectively as follows:

$$\mu_2{}^*(d) := 2^{-1/2}3^{1/4}d^{1/2} \approx 0.93060d^{1/2},$$
$$\mu_3{}^*(d) := 3^{-1/2}4^{2/6}d^{2/3} \approx 0.91649d^{2/3},$$
$$\mu_4{}^*(d) := 4^{-1/2}5^{3/8}d^{3/4} \approx 0.91429d^{3/4},$$
$$\mu_5{}^*(d) := 5^{-1/2}6^{4/10}d^{4/5} \approx 0.91575d^{4/5},$$
$$\mu_6{}^*(d) := 6^{-1/2}7^{5/12}d^{5/6} \approx 0.91844d^{5/6}.$$

These for $l \ge 3$ notably differ from criterions $\{\lambda_l{}^*(d)\}$ used by Fishman and Moore for their spectral tests.[29] **(B)** The distance $\varepsilon_l{}^*(d) = 2^{1/2}d^{(l-1)/l}(l+1)^{-1/(2l)}$ between nearest neighbor lattice points in the l-dimensional regular lattice $L_l{}^*(d)$ has the following explicit values for respective $l = 2, 3, \cdots, 6$:

$$\varepsilon_2{}^*(d) := 2^{1/2}3^{-1/4}d^{1/2} \approx 1.07457d^{1/2},$$
$$\varepsilon_3{}^*(d) := 2^{1/2}4^{-1/6}d^{2/3} \approx 1.12246d^{2/3},$$
$$\varepsilon_4{}^*(d) := 2^{1/2}5^{-1/8}d^{3/4} \approx 1.15649d^{3/4},$$
$$\varepsilon_5{}^*(d) := 2^{1/2}6^{-1/10}d^{4/5} \approx 1.18222d^{4/5},$$
$$\varepsilon_6{}^*(d) := 2^{1/2}7^{-1/12}d^{5/6} \approx 1.20251d^{5/6}.$$

(C) The l-dimensional regular lattice $L_l{}^*(d)$ has the distance $\varepsilon_l{}^{**}(d)$ between next nearest neighbor lattice points related to the nearest

[29]Fishman-Moore criterions are *theoretically smallest possible distances* of parallel $(l-1)$-dimensional lattice hyperplanes in l-dimensional lattices with volume d^{l-1} for their unit cells:

$$\lambda_2{}^*(d) := 2^{-1/2}3^{1/4}d^{1/2} \approx 0.93060d^{1/2} = \mu_2^*(d),$$
$$\lambda_3{}^*(d) := 2^{-1/6}d^{2/3} \approx 0.89090d^{2/3} < \mu_3^*(d),$$
$$\lambda_4{}^*(d) := 2^{-1/4}d^{3/4} \approx 0.84090d^{3/4} < \mu_4^*(d),$$
$$\lambda_5{}^*(d) := 2^{-3/10}d^{4/5} \approx 0.81225d^{4/5} < \mu_5^*(d),$$
$$\lambda_6{}^*(d) := 2^{-1/2}3^{1/12}d^{5/6} \approx 0.77490d^{5/6} < \mu_6^*(d).$$

neighbor distance $\varepsilon_l{}^*(d)$, as the list below shows their explicit relations:

$$\varepsilon_2{}^*(d) = \gamma_2\varepsilon_2{}^{**}(d), \quad \gamma_2 = (2/6)^{1/2} \approx 0.57735,$$
$$\varepsilon_3{}^*(d) = \gamma_3\varepsilon_3{}^{**}(d), \quad \gamma_3 = (3/8)^{1/2} \approx 0.61237,$$
$$\varepsilon_4{}^*(d) = \gamma_4\varepsilon_4{}^{**}(d), \quad \gamma_4 = (4/10)^{1/2} \approx 0.63246,$$
$$\varepsilon_5{}^*(d) = \gamma_5\varepsilon_5{}^{**}(d), \quad \gamma_5 = (5/12)^{1/2} \approx 0.64550,$$
$$\varepsilon_6{}^*(d) = \gamma_6\varepsilon_6{}^{**}(d), \quad \gamma_6 = (6/14)^{1/2} \approx 0.65465.$$

■

5.4 Spectral Tests on Regular Simplex Criterions

We have found all necessary *regular simplex criterions*. We embark on clear definitions of tests based on them to be performed on MC (d, z) random number generators. We begin with the simplest of them, spectral tests.

The spectral tests compare the largest separation $\mu_l(d, z)$ of the $l-1$-dimensional parallel lattice hyperplanes in the MC (d, z) lattice, $L_l(d, z)$ for $2 \le l \le 6$, with that of $\mu_l{}^*(d)$ of the regular lattice $L_l{}^*(d)$.

Definition 35. (Degree l spectral test on regular simplex criterion) Let (d, z) be an MC generator with the lattice $L_l(d, z)$ of its l consecutive outputs together with the dual lattice $L_l{}^*(d, z)$ and its shortest dual lattice vector $\boldsymbol{f}_l{}^{(\min)}$. Then the MC generator (d, z) passes the l-th degree spectral test in the regular simplex criterion if and only if valuations of spectral tests in the regular simplex criterions

$$1 < \frac{\mu_l(d, z)}{\mu_l^*(d)} = \frac{l^{1/2}d^{1/l}}{(l+1)^{(l-1)/(2l)}\|\boldsymbol{f}_l{}^{(\min)}\|} < 1.25, \quad 2 \le l \le 6$$

hold true. **(End of Definition 35)**

Necessary tasks in spectral tests on the regular simplex criterions are the same search for the shortest vectors by the convenient algorithm of Dieter, as they had been in criterions of Geometry of Numbers. However, valuations can now be less than 1, because regular simplex criterions $\{\mu_l{}^*(d)\}$ for $l \ge 3$ are not the smallest, a drastically different circumstances. We abandon such generators with new valuations below 1 as *not excellent*. Our standpoint is that we should pay much attention to the estrangement of the MC *lattices* $\{L_l(d, z)|\ 3 \le l \le 6\}$ from their regular lattice forms. If the lattice $L_l(d, z)$ of the MC generator has small estrangements from the regular lattice $L_l{}^*(d)$, the clear separation $\{\varepsilon_l{}^*(d)|\ 3 \le l \le 6\}$ from the next nearest neighbor distances will remain unmixed. The $l = 2$-dimensional lattice is an exception; the 2-dimensional regular simplex takes the smallest possible value. Please see **Section 6.1** for the explanation.

In passing, we remind ourselves that in **Section 3.1** we noted 4 base sequences on (d, z), (d, z^{-1}), $(d, -z)$ and $(d, (-z)^{-1})$. The introduction of regular simplex criterions brings us a significant convenience that these base sequences share the identical test valuations in spectral tests, together with in all other tests to be introduced shortly. The circumstance is obvious with the (d, z^{-1}) base sequence which is a simple reverse progression of (d, z) MC sequence. As to MC $(d, -z)$ base sequence the tests are reproduced identically if all odd-numbered coordinate axes of (d, z) lattices are taken in the opposite direction; this possibility was commented on the construction of regular simplexes and lattices. The matter is obviously the same with $(d, (-z)^{-1})$ base sequence.

5.5 Edge Tests on Regular Simplex Criterions of MC Generators

The fundamental prospect to be noted is that departures of MC lattices $L_l(d, z)$ for $2 \le l \le 6$ from regular lattice $L_l{}^*(d)$ are to form continuous processes geometrically, and we have possibilities to find excellent MC generators for which lattice deformations remain in small vicinities of regular lattice configuration. In particular, we have the clear separation of nearest neighbor distance $\varepsilon_l{}^*$ and the next nearest distance $\varepsilon_l{}^{**}$ in regular lattices; this separation will persist if the deformation is small. We take that this circumstance is our chance for tests to select excellent MC (d, z) random number generators. Note that $l + 1$ vertices formed by consecutive l MC outputs form an l simplex with

$$_{l+1}C_2 = (l+1)!/\{2!(l-1)!\} = (l+1)l/2$$

edges. Before deformation edges of a regular simplex give the shortest vectors (edges) with one and the same length. If the deformation is small, they will still constitute the same number of lattice vectors forming the shortest-length group in the lattice $L_l(d, z)$.

This comprehension leads us to a powerful new definition of *edge tests*.

Definition 36. (Edge tests on regular simplex criterions) Let an MC (d, z) generator have the lattice $L_l(d, z)$ formed by its consecutive l outputs. Let $N(l) = (l+1)l/2$ shortest lattice vectors of $L_l(d, z)$ be $\{\mathbf{e}_1, \mathbf{e}_2, \cdots, \mathbf{e}_{N(l)}\}$ satisfying conditions:

(A) The set $\{\mathbf{e}_1, \mathbf{e}_2, \cdots, \mathbf{e}_l\}$ are linearly independent.
(B) There holds

$$\varepsilon_l{}^*(d)/1.25 < \|\mathbf{e}_1\| \le \|\mathbf{e}_2\| \le \cdots \le \|\mathbf{e}_{N(l)}\| < \varepsilon_l{}^*(d)/0.67.$$

If these criterions are fulfilled for $2 \le l \le 6$, we define that the MC generator (d, z) pass edge tests.[30]

(End of Definition 36 of edge tests)

The criterions of edge tests introduced above seem to suggest too laborious computing. We are nevertheless able to report on the discovery and excellent existence of MC generators passing these criterions with flags high. Let us keep this know-how as the secret of the HRF, for the coming disclosures on (d, z) in **Chapter 7** will enable you to confirm resulting fine properties straightforwardly, though after a week or two of computing. Please think over *years* of computing that we in fact needed to find reported passers.

A notable point is this. Based on criterions borrowed from *Geometry of Numbers*, nobody could construct an excellent *composite* MC passer $(d = p_1 p_2, z)$ from excellent passers (p_1, z_1) and (p_2, z_2) combined by Sunzi's theorem. This was a puzzling mystery for a

[30] We may more precisely state that the (d, z) MC sequence pass the longest and the shortest edge tests.

long time. Circumstances tumbled by regular simplex criterions. We have already found 2 sets of excellent passer MC generators (and 8 associated base sequences) with composite moduluses formed by odd primes *on regular simplex criterions*. Please see Chapter 7. This stern fact proves that regular simplex criterions should be the true existence with no room for alternatives.

Chapter 6

Extended Second Degree Tests

6.1 Second Degree Tests Revisited

All tests discussed so far give necessary conditions for the statistical excellence of MC generators. Tests have respective computational loads which increase very rapidly with the degree l, the number of consecutive outputs to be tested. Second-degree tests among them

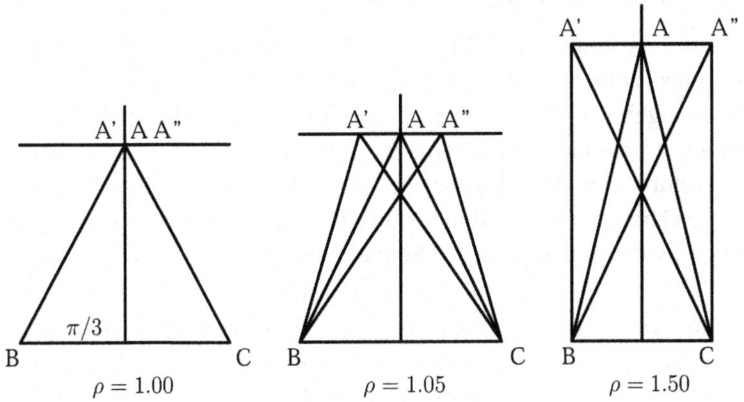

Figure 4 Deformations of a triangle keeping the area.

Random Number Generator on Computers
Naoya Nakazawa and Hiroshi Nakazawa

ISBN 978-981-4968-49-2 (Hardcover), 978-1-003-41060-7 (eBook)
www.jennystanford.com

are the most facile, as requiring only very small computing time. Yet their wise use gives significant first step to reduce the whole test procedures. This should fully be appreciated here, as the significant technique of *extended 2nd-degree tests.*

Corollary 37. In 2-dimensional MC lattice $L_2(d, z)$, the smallest value of the largest distance between parallel and neighboring lattice lines is realized by the case of the regular lattice $L_2{}^*(d)$.

(Proof) Consider **Figure 4** above showing triangles with the same area. The vertex A' or A'' are assumed to give the largest height in the triangle facing the shortest base edge line $\overline{BC}$. With the area of the triangle kept constant, the vertex A' or A'' may be moved parallel to $\overline{BC}$ to the vertex A forming an isosceles ΔABC. Then we lower the vertex A keeping the isosceles form and keeping the area by enlarging the base line $\overline{BC}$. The vertex A keeps its largest height status in the triangle, until the regular form is realized. Thus, the shortest height of the largest distance between parallel and neighboring lattice lines is realized by a regular triangle, or by the regular 2-simplex, or in the regular 2-dimensional triangular lattice $L_l{}^*(d)$ in physical terms. ■

For dimensions $l \geq 3$ **Theorem 34 (A)** shows that the regular simplex criterions of the largest distances $\mu_l{}^*(d)$ between parallel and neighboring lattice hyperplanes in $L_l{}^*(d)$ are larger than the smallest possible values $\lambda_l{}^*(d)$ given by the Geometry of Numbers. Therefore, some MC (d, z) lattices in fact give their largest distances $\mu_l(d, z)$ between $(l-1)$-dimensional parallel and neighboring lattice hyperplanes in the range of **Theorem 34(A)**:

$$\lambda_l{}^*(d) < \mu_l(d, z) < \mu_l{}^*(d).$$

As said previously, this fact makes no problem in technological practices; simplex tests and others invariably prove that such MC generators are not excellent. **Corollary 37** noted above implies that in dimension 2 this complication does not arise at all.

A clearer comprehension of circumstances in dimension 2 is possible to the following effect.

Corollary 38. (Equivalence of 2-dimensional spectral and shortest-edge tests) An MC (d, z) generator has the following 2nd-degree spectral test criterion for the largest distance $\mu_2(d, z)$ of parallel and neighboring lattice lines in $L_2(d, z)$ compared to $\mu_2{}^*(d)$ of the regular lattice $L_2{}^*(d)$:

$$1 < \frac{\mu_2(d, z)}{\mu_2{}^*(d)} < 1.25.$$

The same generator has the 2nd-degree criterion in shortest-edge tests:

$$1 < \frac{\text{nearest neighbor distance } \varepsilon_2{}^*(d) \text{ in } L_2{}^*(d)}{\text{length } \varepsilon_2(d, z) \text{ of the shortest edge in } L_2(d, z)} < 1.25.$$

These criterions agree and both tests are equivalent.

(Proof) In 2-dimension we do not need to go to lattices. Let regular MC lattice $L_2{}^*(d)$ have a regular 2-simplex, spanned by a set of basis vectors, with the edge length $s^* = \varepsilon_2{}^*(d)$ and the height of the vertex $h^* = \mu_2{}^*(d)$. The deformed lattice $L_2(d, z)$ has a 2-simplex corresponding to spanned basis vectors with the largest height $h = \mu_2(d, z)$ and the facing shortest edge of length $s = \varepsilon_2(d, z)$. The same area condition gives $h^*s^* = hs = d$ or $h/h^* = s^*/s$. Thus, the spectral test valuation h/h^* is the same as the (shortest) edge test valuation s^*/s. ■

Hence in 2-dimension only one of spectral test or shortest edge test is necessary. We shall continue to use 2-dimensional spectral test with dual lattice methods of Dieter.

6.2 Extended (Generalized) Second Degree Spectral Tests

In order to realize shorter computing time, the schedule of tests of MC generators should be planned as follows:

(1) 2nd-degree spectral tests.

(2) 3rd-degree spectral tests, and longest as well as shortest edge tests.

(3) 4th-degree spectral tests, and longest as well as shortest edge tests.

(4) 5th-degree spectral tests, and longest as well as shortest edge tests.

(5) 6th-degree spectral tests, and longest as well as shortest edge tests.

These realize a set of tests which are the strictest ever on MC (d, z) generators. We are happy to stress that they have now rich passers. Their marvelous power will be seen in **Chapters 7–9**. We present here the remaining accounts on the 2nd-degree tests and *extended 2nd-degree (spectral) tests*, which form very fast and efficient sieves to reject unqualified generators from laborious considerations. The technique was patented, and now IS presented to THE scientific world for free use.[31]

The MC (d, z, n) random number sequence is $\{nz^0, nz^1, nz^2, \cdots\}$. Taken every $k \geq 1$ steps, they are

$$\{nz^j, nz^{j+k}, nz^{j+2k}, \cdots\} = nz^j\{z^0, z^k, z^{2k}, \cdots\}.$$

This is the MC sequence emitted from the generator (d, z^k, nz^j). In order for this sequence to be excellent, it will be *necessary* that all these generators for $k = 1, 2, \cdots$ are excellent. But we cannot wish that much. At present we shall at best be admitted to wish for 6 consecutive outputs to appear independent. Within this range we may perform swift 2nd-degree tests on generators

$$(d, z) \text{ for all of consecutive outputs} : nz^j\{z^0, z^1, z^2, z^3, z^4, z^5, \cdots\},$$

$$(d, z^2) \text{ outputs with 2-steps apart} : nz^j\{z^0, z^2, z^4, \cdots\},$$

$$(d, z^3) \text{ outputs with 3-steps apart} : nz^j\{z^0, z^3, \cdots\},$$

$$(d, z^4) \text{ outputs with 4-steps apart} : nz^j\{z^0, z^4, \cdots\},$$

$$(d, z^5) \text{ outputs with 5-steps apart} : nz^j\{z^0, z^5, \cdots\}.$$

We wish to have the appearance of independence of consecutive 2-tuples in these sequences. The method to select them is obvious by now and is given in the following.

---

[31]The extended 2nd-degree tests were invented in August 2014 by Naoya Nakazawa and Hiroshi Nakazawa, in *Constructive design of uniform and independent random number generators*, and was granted Russian patent No. 2583729 and U.S. patent No. 2778913 under the title *The method of generalized 2nd-degree spectral tests*. Please see the linked URL *www10.plala.or.jp/h-nkzw/indexarchive20jan1.html.*

Theorem 39. (Extended or generalized 2nd-degree spectral tests) The MC generator (d, z) is defined to pass the extended 2nd-degree spectral tests up to the 5-th power, if valuations $\rho_2(d, z^k)$, namely the ratio $\rho_2(d, z^k) = \mu_2(d, z^k)/\mu_2{}^*(d)$ of

$$\mu_2(d, z^k) := \text{the largest distance of parallel neighboring lattice lines in the lattice } L_2(d, z^k) \text{ and}$$

$$\mu_2{}^*(d) := \text{the largest distance of parallel neighboring lattice lines of the regular lattice } L_2{}^*(d)$$

satisfies $1 < \mu_2(d, z^k) < 1.25$ for $1 \leq k \leq 5$. The restriction takes the form below in terms of the shortest dual lattice vector $\boldsymbol{f}_{\min}(d, z^k)$:

$$1 < \rho_2(d, z^k) = 2^{1/2}3^{-1/4}d^{1/2}/\|\boldsymbol{f}_{\min}(d, z^k)\| < 1.25, \quad 1 \leq k \leq 5.$$

(Proof) The 2-dimensional shortest non-zero dual lattice vector $\boldsymbol{f}_{\min}(d, z^k)$ gives $\mu_2(d, z^k) = d/\|\boldsymbol{f}(d, z^k)\|$ for the largest distance between neighboring lattice lines, while the regular 2-simplex gives $\mu_2{}^*(d) = 2^{-1/2}3^{1/4}d^{1/2}$ for it. By **Corollary 37** applied on the lattice $L_2(d, z^k)$, $\mu_2(d, z^k)$ for integer lattices cannot reach $\mu_2{}^*(d)$ requiring irrational coordinates which realizes the smallest in the 2-dimension. Thus, the criterion of the extended 2nd-degree spectral test of (d, z^k) is obtained as

$$1 < \frac{\mu_2(d, z^k)}{\mu_2{}^*(d)} = 2^{1/2}3^{-1/4}d^{1/2}/\|\boldsymbol{f}_{\min}(d, z^k)\| < 1.25,$$

for $1 \leq k \leq 5$. ■

In passing we should comment on the notion that 2 random numbers a, b are *correlation-free*. If we express the average as $< \cdots >$, we say that a and b are correlation-free if the equality

$$< ab > - < a >< b > = < (a- < a >)(b- < b >) > = 0$$

holds true. If this equation is approximately true, then a and b are nearly correlation free. These matters expressed in the language of probability and averages are very hard to be expressed for MC random numbers. What can be done best will be to rephrase that 2 MC random numbers nz^i and nz^j emitted from the MC (d, z^{j-i}, n)

generator form a nearly triangular lattice; then MC 2-tuples have a close disguise to be correlation free. This language of the world of MC (d, z, n) random numbers cannot have the same meaning as in the above-noted sense of the probability space based on real numbers and of the measured theoretical probability. As Kronecker stated, God made numbers and all the rest is the work of man. MC Random numbers reside in the world of numbers and do not know the work of man. Random numbers on computers brought us to this pre-historic world, but of course there are ways to understand them, though circuitously. We thus continue to use the terminology *(nearly) correlation-free.*

Chapter 7

Three Excellent MC Generators

Introduced criterions for excellent MC generators form an intricate system. To see their actual performances, we show here three excellent MC generators so far obtained in nearly 5 years. Please feel how rarely we could find excellent MC generators that pass all necessary criterions, sympathizing how beautiful they are.

7.1 The Present Best MC Generator #001

Our luck was that the most excellent MC generator #001 was found first of all by Naoya Nakazawa in November, 2018. It employs the modulus formed by 2 SG primes giving a large period about 2^{52}. Constitution is as follows.

Random Number Generator on Computers
Naoya Nakazawa and Hiroshi Nakazawa

ISBN 978-981-4968-49-2 (Hardcover), 978-1-003-41060-7 (eBook)
www.jennystanford.com

$d = p_1 p_2 = 18055400005099021 \approx 2^{54.00}$
SG prime $p_1 = 134265023 \approx 2^{27.00}$
SG prime $p_2 = 134475827 \approx 2^{27.00}$
the primitive root z_1 of p_1: $z_1 = 19061252 \approx 2^{24.18}$
the primitive root z_2 of p_2: $z_2 = 77600525 \approx 2^{26.21}$
composite multiplier by Sunzi theorem:

$$z \equiv 7759097958782935 \approx 2^{52.79}$$

usable LCM period T:

$$T = \mathrm{LCM}(p_1 - 1, p_2 - 1)/2 = 4513849934089543 \approx 2^{52.00}$$

Multipliers sharing usable period and test valuations $(\mathrm{mod}\ d)$:

$$\begin{aligned} z^{-1} &\equiv 8723774547862110 \approx 2^{52.95} \\ -z &\equiv 10296302046316086 \approx 2^{53.19} \\ -z^{-1} &\equiv 9331625457236911 \approx 2^{53.05} \end{aligned}$$

Figure 5 The constitution of #001.

We show below valuations of #001 in tests.

Test Valuations A. Valuations of extended 2nd tests for MC generators $(d, z^1), (d, z^2), \cdots, (d, z^{11})$:

1.08678338 1.23476055 1.09373237 1.14778981 1.13682785
1.16390618 1.09784908 1.21656428 1.52552804 1.34934813
7.69460527

Test Valuations B. Valuations of regular simplex spectral tests for degree 3 to degree 6:

1.13600074 1.04031015 1.10996227 1.21389160

Test Valuations C. Longest and shortest edge test valuations of 3rd to 6th degrees:

(3rd degree)	0.78489424	1.18938572
(4th degree)	0.73780699	1.17913686
(5th degree)	0.83524952	1.20173353
(6th degree)	0.71002135	1.20574247

Conveniences given by the modulus $d = p_1 p_2$ are worthy of special mention. Since primes p_1 and p_2 are in integer*4 class, integer subrandom numbers may be obtained and stored wholly in the integer*8 arithmetic. Moreover, combined MC random numbers enjoy a marvelous structure of Sunzi's theorem, as found remarkably by Naoya Nakazawa. See the computing program in the **Figure 6** on the next page that enables us to compute all within integer*8 arithmetic only.

Valuations of tests are all excellent with this #001. In particular, the results of the longest edge tests are splendid. The (d, z) lattice of #001 has its original regular simplex deformed in such a tame way that the longest edges does not reach even the next nearest neighbor distance in regular the lattice case. This excellence also gives significant theoretical information. In the 20th-century spectral tests with criterions borrowed from the geometry of numbers, a number of passer MC generators, (p, z)'s for prime-primitive root pairs, were certainly found. However, when those passers were combined by Sunzi's theorem to a 2-prime modulus $d = p_1 p_2$, nobody could find a passer on the same criterions. The authors were really perplexed by these results against our intuition, that excellent subgenerators should have a larger chance to be combined to a good generator. This difficulty was the motive that drove the authors to introduce new regular simplex criterions. The discovery of #001 proved that new criterions are correct and fertile enabling us to have extraordinarily high speeds of computing. See **Figure 6**, in which we give a FORTRAN subprogram of #001 to generate consecutive 100 random numbers.

The program proceeds only in integer*8 and real*8 arithmetic. Please feel the power of the **Sunzi's theorem**.[32] The CPU time for these 10 million outputs is about 0.281s on our middle-speed computer. Random number outputs have the sufficient $1/d \approx 2^{-54}$ precision for real*8. The usable period of the generator is about 2^{52}.

[32]This implication of Sunzi's theorem was noticed first by Naoya Nakazawa. Readers are referred to Naoya Nakazawa and Hiroshi Nakazawa: English translation of a Patent Application to Japan Patent Office, June 13 of 2022, *Sunzi Reduction to Compute Multiplicative Congruential Random Numbers with Composite Moduluses*, in *http://www10.plala.or.jp/h-nkzw/indexarchive22june13.html.*

Consecutive generation of its whole 2^{52} sequence will need 2 years on our desktop computers. For the convenience of readers, we show further below in **Figure 7** 100 MC random number outputs after the end of noted 10^7 outputs computed by the program of **Figure 6**.

```
program main
implicit integer*8(i-n), real*8(a-h,o-z)
common ip1,ip2,id,ad,iz1,iz2,mz1,mz2,ip2mp1,ip1mp2
ip1=134265023
ip2=134475827
id=ip1*ip2   ! id ≈ 2^54
ad=id
iz1=19061252
iz2=77600525
n1=10
n2=13
ip2mp1=52577007
ip1mp2=81816271
iseed1=n1 ! integer*4 to assign iseed1=mod(iseed,ip1)
iseed2=n2 ! integer*4 to assign iseed2=mod(iseed,ip2)
mz1=iseed1
mz2=iseed2
do i=1,10000000
call random(rand)
end do
      ..................
end   !(main program end)

subroutine random(rand)
implicit integer*8(i-n), real*8(a-h,o-z)
common ip1,ip2,id,ad,iz1,iz2,mz1,mz2,ip2mp1,ip1mp2
mz1=mod(mz1*iz1,ip1)
mz2=mod(mz2*iz2,ip2)
mz1a=mod(mz1*ip2mp1,ip1)
mz2a=mod(mz2*ip1mp2,ip2)
az=mod(ip2*mz1a+ip1*mz2a,id)
rand=az/ad
return
end
```

Figure 6 Computing Programs for #001.

0.653816355434	0.162395903492	0.666319058508	0.192823573723
0.489788498203	0.327381692216	0.006207372410	0.817190447249
0.639382522876	0.999243182851	0.717807517328	0.582888069563
0.751959446280	0.456610909409	0.201265518413	0.197352588136
0.185468833692	0.026325012527	0.798951425190	0.980168183205
0.728774197785	0.895636674003	0.746279846438	0.334966215203
0.163132201425	0.161807776678	0.478463418819	0.402555313497
0.412925462471	0.228549325709	0.116935094385	0.887686052660
0.748053624507	0.372387517800	0.401887611920	0.513438563398
0.218008135464	0.479107340785	0.371799991246	0.610473874869
0.495998588197	0.704020149239	0.125946074052	0.689497113384
0.296979898817	0.664141855353	0.967378082658	0.861373665256
0.146986132091	0.320681156594	0.293103638455	0.410906693576
0.830103955630	0.320270139018	0.924042828676	0.087350373543
0.943936287549	0.994736329597	0.408045545191	0.472523592537
0.765124568761	0.630798202897	0.873021775270	0.612753906188
0.124335863299	0.816173368849	0.470611672694	0.880574814564
0.668048788679	0.380757635873	0.439593072176	0.532108985606
0.235689661806	0.023857729188	0.606829757993	0.762549693572
0.627268755976	0.592982603016	0.803692768860	0.555192595035
0.568754093418	0.482014270564	0.450138517310	0.960974827043
0.500236510179	0.285096197971	0.920893782638	0.842851188064
0.083495098650	0.555523403307	0.712500499476	0.525529497885
0.921528283135	0.667901000917	0.272780491599	0.962922804725
0.924506422608	0.495041819614	0.783468131560	0.851983710989

Figure 7 100 Outputs from #001 generator.

7.2 Prime-Primitive Root Generator #M001

The second discovery by Naoya Nakazawa in February 2019 was a prime-primitive root generator (p, z) #M001 needing integer*8 treatments. The constructions and performances are shown below.

$d = p =$ NN prime17179869989 $\approx 2^{34.00}$
the primitive root multiplier $z = 7928410072 \approx 2^{32.88}$
usable period: $T = (p-1)/2 \approx 2^{33.00}$.
the period and test valuations are shared by 3 multipliers:

$$\begin{aligned} -z &\equiv 9251459917 \approx 2^{33.11}, \\ z^{-1} &\equiv 7620680202 \approx 2^{32.83}, \\ -z^{-1} &\equiv 9559189787 \approx 2^{33.15}. \end{aligned}$$

Test Valuations A. Valuations[33] of extended 2nd tests for MC generators $(p, z^1), (p, z^2), \cdots, (p, z^{11})$:

1.18541312 1.13773700 1.10002050 1.11668346 1.16206454
1.46032715 1.28536419 1.18985913 1.30109005 1.58479668
2.69897538

Test Valuations B. Valuations of regular simplex spectral tests for degree 3 to degree 6:

1.14596592 1.09599706 1.03775043 1.17623397

Test Valuations C. Longest and shortest edge test valuations of 3rd to 6th degrees:

(3rd degree)	0.80861627	1.12182344
(4th degree)	0.71069728	1.24592478
(5th degree)	0.73916415	1.17581188
(6th degree)	0.74510834	1.18352890

Figure 8 The small-scale MC generator #M001.

These valuations are beautiful. However, too many difficulties are to be reported on this generator.

(1) The usable period $T \approx 2^{33.00}$ is too short for simulational use of today. The whole period T will be used up in about 6 minutes if the computer is used to generate these MC random numbers only.

(2) The modulus $d = p$ and the multiplier z are of integer*8 characters. Yet the output random numbers may be used only for real*4 precision. Moreover, we need to use real*16 variables in computing multiplication by z and in taking modulo p. This necessity is too much for this small real*4 MC random numbers on MC #M001 generator. Below we

[33]In relation to **Chapter 8** we show up to (p, z^{11}).

show a short FORTRAN program to compute 10^7 random numbers. It will convince you decisively that #M001 is not a practical MC generator.

```
program main
implicit integer*8(i-n), real*8(a-g,o-p,r-z), real*16(q)
ip=17179869989d0   ! the prime ip is about 2^34.00
rp=ip
iz=7928410072d0
in=10   ! the seed is 10
qmz=in
do i=1, 10000000
qmz=mod(qmz*iz,rp)
rand=mod(qmz,rp)/rp
end do
end
```

Figure 9 Program on #M001 for 10^7 outputs.

The CPU time required for this #M001 generator is 2.531s, about 10 times larger than that for #001. This fact decisively disproves the practicability of #M001.

7.3 A Large-Scale Excellent MC Generator #003

The generator #003 was found by Naoya Nakazawa in April 2020. Please measure up how rarely good MC random number generators could be found.[34] Its modulus d is formed by 2 NN primes p_1, p_2 as $d = p_1 p_2$. Test valuations are almost comparably beautiful as #001, but please see the generalized 2nd tests. The Usable LCM period

[34]We should let 10 or more cores of computers run 24 hours without rest, consuming sizable electric power.

is $T \approx 2^{51}$ which is a little smaller than #001, as subgenerators (p_1, z_1) and (p_2, z_2) both have the common factor 4 in their usable periods. Putting aside this small drawback, #003 will be a good alternative of #001, with the following structures.

Modulus $d = p_1 p_2 = 18015370515269401 \approx 2^{54.000}$
NN subprime $p_1 = 134224829 \approx 2^{27.000}$
NN subprime $p_2 = 134217869 \approx 2^{27.000}$
primitive root submultiplier $z_1 = 95967890 \approx 2^{26.516}$ of p_1
primitive root submultiplier $z_2 = 4256141 \approx 2^{22.021}$ of p_2
Sunzi multiplier $z = 16048994718289548 \approx 2^{53.833}$
usable period $T = 2251921280853338 \approx 2^{51.000}$
multipliers sharing usable period and test valuations

$$\begin{aligned} z^{-1} &\equiv 10990185200333827 \approx 2^{53.29} \\ -z &\equiv 1966375796979853 \approx 2^{50.80} \\ -z^{-1} &\equiv 7025185314935574 \approx 2^{52.64} \end{aligned}$$

Figure 10 Generator #003 on 2 NN prime modulus.

Statistical test valuations of #003 are almost comparable as #001. If we need a generator other than #001 for the confirmations of simulational results, this generator will provide a very good alternative.

Test Valuations A. Generalized 2nd-Degree tests of (d, z^j) for $1 \leq j \leq 11$:

1.12378644 1.22759925 1.15381455 1.07582363 1.12113014 1.90830600 2.56595210 1.64729694 1.10578807 1.10728840 2.12669792

Test Valuations B. 3-6th Spectral Test Valuations in Regular Simplex Criterions:

1.14537815 1.06716995 1.13487872 1.21563615

Test Valuations C. Longest and Shortest Edge Test Valuations of 3rd to 6th Degrees:

(3rd Degree)	0.77772641	1.16750024
(4th Degree)	0.74018574	1.20907497
(5th Degree)	0.68729723	1.23300972
(6th Degree)	0.69782364	1.23425488

Valuations are excellent. The generalized 2nd-degree tests pass up to the 5th degree, which is a little behind the 8th of #001. Thus, if used as distributed on time-space lattice, a lattice point may at the maximum be distributed of $\rho = 3$ independent random numbers. Compare this value to $\rho = 4$ of #001 described in **Section 7.1**.

Chapter 8

Distribution of MC Random Numbers on Spatial Lattices

8.1 Distribution of MC Random Numbers on Spatial Lattices

In 1992 Ferrenberg, Landau and Wong[35] raised a problem to distribute random initial values to Ising spin magnetic moments on 2-dimensional lattice, aiming to compare the results to famous exact solutions for the phase-transition, using several random number generators which were then prosperous. They reported sizable errors between the theory and simulations, and concluded as they titled. Though the report was not very clear how they used random numbers in their analysis as initial values and/or in some Ising dynamics.

The random number systems used at their time are well guessed to have many problems in *independence*, as the criterions used then were not correct. What we can now infer is: *What will come about if we succeed in using random numbers with clearer independence*

[35]M. Ferrenberg, D. P. Landau and Y. J. Wong, *Monte Carlo simulation: Hidden errors from 'good' random number generators, Physical Review Letters* **69** (1992), pp. 3382–3384.

Random Number Generator on Computers
Naoya Nakazawa and Hiroshi Nakazawa

ISBN 978-981-4968-49-2 (Hardcover), 978-1-003-41060-7 (eBook)
www.jennystanford.com

or with less prominent correlation compared to those distributed on neighboring spins. We imagine that the uniformity of random numbers at that time might well pass the level of our present eyes.

We present here a new way to distribute MC random numbers on an s-dimensional spatial lattice $L := J_1 \times J_2 \times \cdots \times J_s$ with integers $J_1, J_2, \cdots, J_s$. The origin of the discrepancy in tests of Ferrenberg et al. seems to be clear. Multiplicative congruential (MC) random integers are emitted by the recursive congruence relation *sequentially* as

$$z_0 := n \pmod{d}, \quad z_i \equiv zz_{i-1} \equiv nz^i \pmod{d}, \quad i = 1, 2, 3, \cdots .$$

We turn them into a *sequence* of rational numbers

$$\{x_i := z_i/d \mid 0 < x_i < 1, \quad i = 0, 1, 2, \cdots\},$$

and use geometrical properties of the related lattice to find a good disguise of independence for consecutive 6 random numbers. Now we face problems in distributing MC random numbers on 2- or higher dimensional spatial lattices. We are out of the reign of **Theorem 6** here. Yet we now have generalized 2nd tests. Let us see what happens with a tentative distribution.

Figure 11 shows generalized 2nd-degree test valuations of MC random numbers $\{x_j \pmod{d} \mid 1 \le j \le 100\}$ distributed in 5×20 array. Results indicate that MC random integers nz^i and nz^j may retain strong correlations however large $|j - i|$ might be. Yet there exist certain consecutive groups of outputs that seem to keep a good disguise of non-correlation. The matter looks to be the problem of *search*. We present here the results of our efforts in this suggested direction.

The conceptual ideal of the random initial value problem posed in the problem of Ferrenberg et al. is inferred clearly. If random numbers with uniformity and perfect mutual independence are available for any length, they may be distributed in an arbitrary manner on spatial lattice sites (or particles residing on spatial lattice points) keeping excellent statistics. Yet we should recall that even if (d_1, z_1, n_1) and (d_2, z_2, n_2) are excellent MC generators, their arbitrary combination by Sunzi's theorem to $(d = d_1d_2, z, n)$ can be excellent with a hopelessly small probability. This experience will be true universally. We should take excellent generators, should

1.08678338	1.23476055	1.09373237	1.14778981	1.13682785
1.16390618	1.09784908	1.21656428	1.52552804	1.34934813
7.69460527	2.55311256	1.07606160	1.14834672	1.79878519
1.13707473	1.18702892	1.18798077	2.29260937	1.26015847
1.89030717	1.51673051	3.43169297	1.36133178	4.18359045
1.08360526	1.23562664	1.67564981	1.92069320	1.72402921
1.50425323	1.53280617	1.10976358	1.12535134	1.22395837
1.10127670	1.15054180	1.12720961	2.21855027	2.52780968
3.56247191	1.06371152	1.06812115	1.35886246	2.37481693
3.26454017	1.77787879	2.36950022	1.21949430	1.34171928
2.08445663	3.78953455	2.24975187	3.04568518	2.33680473
1.32609094	1.02518982	1.08406145	1.35721362	1.60367053
5.49139306	2.89263636	2.97450317	1.27366678	1.18959291
1.22942355	1.13900989	1.74356681	1.35042470	1.08121223
1.31745097	2.18173831	1.19013639	1.12329646	1.49539173
1.31313325	1.11606237	1.09933883	1.36108464	1.61520260
2.70842792	1.84275459	1.29922809	1.20642859	1.42634011
1.23265681	1.12422937	4.25875962	5.67068589	3.56911742
1.08253679	3.33218802	1.54819932	1.40667274	6.44602354
1.54269988	1.37483596	1.83651405	11.08455046	1.66059092

Figure 11 Valuations of extended 2nd tests. of (d, z^j) MC #001 Outputs for $1 \leq j \leq 100$

combine them, and use the resultant generator *after tests*, not on simple guesses without tests.

This discrepancy with our intuition should also be mentioned on periodic boundary conditions. If random numbers in an MC sequence show true independence among themselves, ways of their distribution on any lattices give no problems, irrespective of whether the lattice forms a torus or not. In reality, we have to work to realize their adequate distributions, so as to make spatially neighboring MC random numbers to have *the appearance* of mutual independence, or at least the appearance of no correlation, across the periodic boundary.

We report here that MC sequences as introduced hide versatile structures that work nicely on these lattice geometries. The first clue is that any 2 random integers, as emitted from *one and the same* MC generator (d, z, n) give the exclusive form nz^i to all random integers modulo d. Consider arbitrary 2 MC random integers, nz^j and nz^i. The former is the first MC random number emitted from (d, z^{j-i}, n') with $n' \equiv nz^i$ (mod$d$). This at once reminds us of another clue, a necessary condition for the excellence, that is, for these MC random numbers to *appear excellently correlation-free*. This saying involves some questionable point, which will be discussed later in more detail. Here, we turn to endeavor to take a lattice and to find on it excellently distributed MC generators for their *relevant* pairs to give excellent (d, z^{j-i}) in their extended 2nd-degree tests.

8.2 Distribution of Random Numbers on Lattices as a Variable Base Representation of Integers

An important requirement arising in common to all simulations on spatial lattices is that a degree of freedom on a lattice point invariably demands $r \geq 2$ MC random numbers with the *independent* appearance. Of course, $r \leq 6$ is necessary in our present circumstances and relevant MC random numbers should be taken as consecutive outputs of the generator. We thus take invariably an r-component row vector

$$\boldsymbol{v}(r) :\equiv (1, z, z^2, \cdots, z^{r-1}) \ (\text{mod } d),$$

formed by first r elements from the MC generator $(d, z, 1)$. We premise that the distribution of random numbers on points of the spatial lattice is treated always in this form of r-consecutive cluster as the unit.

Lattice points of the s-dimensional integer lattice $G(\boldsymbol{j})$ are denoted roughly as $G(\boldsymbol{j}) = J_1 \times J_2 \times \cdots \times J_s$; a more precise definition is

$$G(\boldsymbol{j}) := \{\boldsymbol{j} = (j_1, j_2, \cdots, j_s) | \ 0 \leq j_i \leq J_i - 1, \quad 1 \leq i \leq s\}.$$

The first basic procedure is to tear the MC random integer sequence into a conjunction of length J_0 pieces. The length J_0 is assumed $J_0 > r$. And we simplify hereafter that the MC sequence is the whole of the cyclic sequence with the infinite length. The first r

element of a J_0-piece thus forms a modified r-vector, $n\mathbf{v}(r)$ (mod d) as multiplied modulo d by the first element n of the piece. Thus, J_0 pieces, in total $J_1 J_2 \cdots J_s$ in number, are concatenated to form the (d, z, n) MC random number vector of the length $J_0 J_1 J_2 \cdots J_s$. Since the first r components of the J_0-piece are included in J_0 piece, $J_0 > r$ is true. The first elements of J_0 pieces are

$$\{nz^{lJ_0} \text{ (mod } d) |\, l = 0, 1, 2, \cdots, J_1 J_2 \cdots J_s - 1\},$$

which are named *roots*. The magnitude of the *stagger* integer J_0 for roots is to be determined by tuning tests. The aggregate of *roots* should exist just as many as lattice points of $G(\boldsymbol{j})$. Roots are placed or *sown* on lattice points of $G(\boldsymbol{j})$, *in an appropriate manner to be soon discussed*. Finally, the length r vector $\mathbf{v}(r)$, the first r components of the length J_0 piece, are grafted on every distributed root. These grafted r MC random numbers are taken to form the all of MC random numbers on the lattice $G(\boldsymbol{j})$, ignoring $J_0 - r$ random numbers are left not used.

We simplify the setting by defining the MC sequence to infinite length

$$(d, z, n) \text{ MC sequence} := \{nz^j \text{ (mod } d) |\, 0 \leq j \leq \infty\}.$$

And we define on each of these infinite sequence its *usable period* T as the smallest $T > 0$ that gives

$$z^T \equiv d - 1 \equiv -1 \text{ (mod } d),$$

or as the period itself if the MC $(d, z, 1)$ cyclic sequence does not have -1 (mod d). We assume that the whole of the length J_0 pieces distributed on all lattice points (which total to $J_1 J_2 \cdots J_s$) satisfy the restriction

$$J_0 J_1 J_2 \cdots J_s < T.$$

This ensures that all distributed MC random numbers are taken only once within a usable period T, and there is no duplicated use of MC random numbers on the lattice $J_1 \times J_2 \times \cdots \times J_s$. All tuned examples disclosed in later sections obey this restriction.

The clue of the way to distribute thus prepared MC random numbers on a spatial lattice is the following.

Corollary 40. (Variable base representation of integers) Let integers $J_1, J_2, \cdots, J_s$ be not less than 2. Corresponding to the representation of an integer K with a fixed base $J \geq 2$,

$$K = j_1 + j_2 J + j_3 J^2 + \cdots, \quad 0 \leq j_i \leq J - 1, \quad i = 1, 2, \cdots,$$

any integer K in the range $0 \leq K \leq J_1 J_2 \cdots J_s - 1$ has its unique representation $K = k(\boldsymbol{j})$ by integers $\boldsymbol{j} = (j_1, j_2, \cdots, j_s)$ with

$$k(\boldsymbol{j}) := j_1 + j_2 J_1 + j_3 J_1 J_2 + \cdots + j_s J_1 J_2 \cdots J_{s-1},$$
$$0 \leq j_i \leq J_i - 1, \quad 1 \leq i \leq s.$$

(Proof) Let $K = k(\boldsymbol{j})$ be defined by integers $\boldsymbol{j} = (j_1, j_2, \cdots, j_s)$ in noted ranges as above. Since $k(\boldsymbol{j}) = k(j_1, j_2, \cdots, j_s)$ is increasing with the increase of components of $\boldsymbol{j}$, we have

$$\begin{aligned} 0 \leq K &\leq (J_1 - 1) + (J_2 - 1)J_1 + (J_3 - 1)J_1 J_2 + \cdots \\ &\quad (J_s - 1)J_1 J_2 \cdots J_{s-1} \\ &= J_1 J_2 \cdots J_{s-1} J_s - 1. \end{aligned}$$

We readily prove that integers $j_1, j_2, \cdots, j_s$ giving the above range of K with $k(\boldsymbol{j})$ are unique: Suppose that another representation of K exists as

$$K = j_1' + j_2' J_1 + j_3' J_1 J_2 + \cdots + j_s' J_1 J_2 \cdots J_{s-1},$$
$$0 \leq j_i' \leq J_i - 1, \quad 1 \leq i \leq s.$$

This implies:

$$0 = (j_1' - j_1) + (j_2' - j_2)J_1 + \cdots + (j_s' - j_s)J_1 J_2 \cdots J_{s-1}.$$

Divide this by J_1. The remainder is 0 of course, which is the remainder of division of $j_1' - j_1$ by J_1. Since $-J_1 < j_1' - j_1 < J_1$ holds true, this remainder 0 is $j_1' - j_1$ itself, which proves $j_1' = j_1$. The quotient Q is

$$Q = 0 = (j_2' - j_2) + (j_3' - j_3)J_2 + \cdots + (j_s' - j_s)J_2 \cdots J_{s-1}.$$

The division of $Q = 0$ by J_2 gives likewise $j_2' - j_2 = 0$ or $j_2' = j_2$. Similar procedures give up to $j_s' = j_s$ obviously. ■

8.3 Random Root Function on Lattices

We have now the first perspective on structures of r-component vectors in length J_0 MC random number pieces as to be distributed on lattice points of $G(\boldsymbol{j})$.

Theorem 41. (Random root function on the lattice) Call integer MC random numbers as the *random root function* and denote as $g(\boldsymbol{j})$ on the lattice point $\boldsymbol{j}$ of $G(\boldsymbol{j})$, in the following way:

$$k(\boldsymbol{j}) := j_1 + j_2 J_1 + j_3 J_1 J_2 + \cdots + j_s J_1 J_2 \cdots J_{s-1},$$

$$\{0 \leq j_k \leq J_k - 1, \quad 1 \leq k \leq s\},$$

$$g(\boldsymbol{j}) :\equiv nz\hat{\ }\{J_0 k(\boldsymbol{j})\} \pmod{d},$$

$$\equiv nz^{j_1 J_0 + j_2 J_0 J_1 + j_3 J_0 J_1 J_2 + \cdots + j_s J_0 J_1 J_2 \cdots J_{s-1}} \pmod{d},$$

$$G(\boldsymbol{j}) := J_1 \times J_2 \times \cdots \times J_s.$$

This function $g(\boldsymbol{j})$ maps random numbers $\{z^{lJ_0} \pmod{d}\}$ or *roots*

$$\{z^{lJ_0} \pmod{d} |\, l = 0, 1, 2, \cdots, J_1 J_2 \cdots J_s - 1\}$$

one to one onto lattice points of $G(\boldsymbol{j})$.

(Proof) The MC generator (d, z, n) emits all distinct random integers within its period included in the usable period T. Since the integer function $k(\boldsymbol{j})$ has totally the number $J_1 J_2 \cdots J_s < T$ of values from 0 up to $J_1 J_2 \cdots J_s - 1$, lattice points of $G(\boldsymbol{j})$ are mapped by *roots* one-to-one onto. The *random root function* $g(\boldsymbol{j})$ is well-defined on $G(\boldsymbol{j})$. ■

8.4 Non-Periodic Tuning of Random Vector Function

Let $\boldsymbol{j}$ be an inner lattice point of $G(\boldsymbol{j}) = J_1 \times J_2 \times \cdots \times J_s$, that is, the lattice point $\boldsymbol{j}$ off the boundary. The nearest $(\boldsymbol{N}_1)$ neighbor lattice points of $\boldsymbol{j}$ to the j_i-axis direction are as follows for $1 \leq i \leq s$:

$$\boldsymbol{j}^{(i+)} := (j_1, j_2, \cdots, j_i + 1, \cdots, j_s),$$

$$\boldsymbol{j} := (j_1, j_2, \cdots, \quad j_i \quad, \cdots, j_s),$$

$$\boldsymbol{j}^{(i-)} := (j_1, j_2, \cdots, j_i - 1, \cdots, j_s).$$

On this inner lattice point $\boldsymbol{j}$ of $G(\boldsymbol{j})$ and on its $\boldsymbol{N}_1$-neighbors, values of random root functions and grafted vectors are as follows:

$$g(\boldsymbol{j}^{(i+)})\boldsymbol{v}(r) :\equiv g(\boldsymbol{j})\boldsymbol{v}(r)z^{J_0J_1\cdots J_{i-1}} \pmod d,$$

$$g(\boldsymbol{j})\boldsymbol{v}(r) \pmod d,$$

$$g(\boldsymbol{j}^{(i-)})\boldsymbol{v}(r) :\equiv g(\boldsymbol{j})\boldsymbol{v}(r)z^{-J_0J_1\cdots J_{i-1}} \pmod d.$$

Dividing off the common MC random number $g(\boldsymbol{j})$, their correlations may be inferred as those of elements of $\boldsymbol{v}(r)$ and $z^{\pm J_0J_1\cdots J_{i-1}}\boldsymbol{v}(r)$.

In order to consider this type of correlations systematically, we prepare a lemma. Let $\boldsymbol{v}(r)^{-1} :\equiv (1, z^{-1}, z^{-2}, \cdots, z^{-(r-1)}) \pmod d$ be the row vector formed by the reverse progression of r-consecutive MC integers.[36] Let also ${}^t\{\boldsymbol{v}(r)^{-1}\}$ denote the column vector, as the transpose of a row vector. We take the matrix product of ${}^t\{\boldsymbol{v}(r)^{-1}\}$ and $z^K\boldsymbol{v}(r)$ modulo d. **Figure 12** gives all relevant multipliers thus formed:

	z^{K+0}	z^{K+1}	z^{K+2}	$\cdots$	z^{K+r-1}
z^0	z^K	z^{K+1}	z^{K+2}	$\cdots$	z^{K+r-1}
z^{-1}	z^{K-1}	z^K	z^{K+1}	$\cdots$	z^{K+r-2}
z^{-2}	z^{K-2}	z^{K-1}	z^K	$\cdots$	z^{K+r-3}
$\cdots$	$\cdots$	$\cdots$	$\cdots$	$\cdots$	$\cdots$
z^{-r+1}	z^{K-r+1}	z^{K-r+2}	z^{K-r+3}	$\cdots$	z^K

Figure 12 Matrix product ${}^t\{\boldsymbol{v}(r)\}^{-1}z^K\boldsymbol{v}(r)$ for $K \geq r$.

Terms $z^{K-r+1}, z^{K-r+2}, \cdots, z^{K-r+r}, z^{K+1}, z^{K+2}, \cdots, z^{K+r-1}$ of **Figure 12** gives the lemma below.

Lemma 42. (MC generators relating $\boldsymbol{v}(r)$ and $z^K\boldsymbol{v}(r)$ for $K \geq r$) Recall $\boldsymbol{v}(r) \equiv (1, z, z^2, \cdots, z^{r-1}) \pmod d$, the starting length

[36]Note that any integer coprime to d has its inverse modulo d, as a member of the reduced residue class group.

r consecutive MC random integers emitted from the (d, z, n) generator, and let K be an integer satisfying $K \geq r$. Correlations of elements of $\boldsymbol{v}(r)$ and elements of $z^K \boldsymbol{v}(r) \pmod{d}$ are exclusively inferred from 2nd-degree tests of $2r - 1$ generators,

$$(d, z^K), \quad (d, z^{K\pm 1}), \quad (d, z^{K\pm 2}), \quad \cdots \quad , \quad (d, z^{K\pm(r-1)}).$$

(Proof) The generator (d, z, n) gives MC random numbers of the form $nz^k \pmod{d}$. We may infer on their correlations taking the common seed n off, and the problem is the correlation of z^k and $z^{k'}$. The matrix form **Figure 12** summarizes all relevant terms. Their distinct forms give the assertion. ■

8.5 Relevant MC Generators in Non-Periodic Tuning for Random Initial Value Problems

The information obtained by the abstract **Lemma 42** is now embodied as tuning procedures of $g(\boldsymbol{j})\boldsymbol{v}(r)$ under non-periodic boundary conditions. On $\boldsymbol{j}$ lattice point of the lattice $G(\boldsymbol{j})$, the

j_1**-axis direction**

$$(\boldsymbol{I}_1) \quad (d, z^{J_0}), \; (d, z^{J_0\pm 1}), \; (d, z^{J_0\pm 2}), \cdots, \; (d, z^{J_0\pm(r-1)}).$$

j_2**-axis direction**

$$(\boldsymbol{I}_2) \quad (d, z^{J_0J_1}), \; (d, z^{J_0J_1\pm 1}), \; (d, z^{J_0J_1\pm 2}), \cdots, \; (d, z^{J_0J_1\pm(r-1)}).$$

j_3**-axis direction**

$$(\boldsymbol{I}_3) \quad (d, z^{J_0J_1J_2}), \; (d, z^{J_0J_1J_2\pm 1}), \; (d, z^{J_0J_1J_2\pm 2}), \cdots, \; (d, z^{J_0J_1J_2\pm(r-1)}).$$

....

j_s**-axis direction**

$$(\boldsymbol{I}_s) \quad (d, z^{J_0J_1J_2\cdots J_{s-1}}), \; (d, z^{J_0J_1J_2\cdots J_{s-1}\pm 1}), \; (d, z^{J_0J_1J_2\cdots J_{s-1}\pm 2}), \; \cdots, \; (d, z^{J_0J_1J_2\cdots J_{s-1}\pm(r-1)}).$$

Figure 13 Matrix product ${}^t\{\boldsymbol{v}(r)\}^{-1} z^K \boldsymbol{v}(r)$ for $K \geq r$.

nearest N_1 neighbors[37] to the j_i-axis direction for $1 \leq i \leq s$ are

$$(j_1, \cdots, j_i, \cdots, j_s) \to (j_1, \cdots, j_i \pm 1, \cdots, j_s), \quad 1 \leq j_i \leq J_i - 2.$$

only *between* inner lattice points, from $j_i = 1$ to $j_i = J_i - 2$. This schematical notation of necessary $J_i - 2$ procedures give **Theorem 43** noted beneath.

Theorem 43. (Generators linking random vector functions on inner lattice points and on their N_1 neighbors) Let (d, z, n) be an MC generator, let $J_0 \geq r$ be true, let $G(\boldsymbol{j}) = J_1 \times J_2 \times \cdots \times J_s$ be the lattice with $J_i \geq 2$ for $1 \leq i \leq s$, let $2 \leq r \leq 6$ be satisfied with $J_0 \geq r$, and let

$$g(\boldsymbol{j}) :\equiv nz\hat{}\{J_0 k(\boldsymbol{j})\} \pmod d$$
$$\equiv nz^{j_1 J_0 + j_2 J_0 J_1 + j_3 J_0 J_1 J_2 + \cdots + j_s J_0 J_1 J_2 \cdots J_{s-1}} \pmod d$$

be the MC (d, z, n) random root function distributed on the lattice $G(\boldsymbol{j})$. To the j_i-direction for any $1 \leq i \leq s$, relevant MC generators that connect elements of the random vector function $g(\boldsymbol{j})\boldsymbol{v}(r)$ at any inner lattice point $\boldsymbol{j}$ and those at N_1 neighboring lattice points, are comprised by $2r - 1$ generators noted in **Figure 13** above:

(Proof) Examining j_i-axis direction for $i = 1, 2, \cdots, s$ with **Lemma 42**, we obtain the assertion. We explain with the case $j_i = j_2$:

$$\boldsymbol{j}^{(2+)} :\equiv (j_1, j_2 + 1, j_3, \cdots, j_s),$$
$$\boldsymbol{j} :\equiv (j_1, j_2, j_3, \cdots, j_s),$$
$$\boldsymbol{j}^{(2-)} :\equiv (j_1, j_2 - 1, j_3, \cdots, j_s),$$

$$g(\boldsymbol{j}^{(2+)})\boldsymbol{v}(r) :\equiv g(\boldsymbol{j})\boldsymbol{v}(r) z^{J_0 J_1},$$
$$g(\boldsymbol{j})\boldsymbol{v}(r),$$
$$g(\boldsymbol{j}^{(2-)})\boldsymbol{v}(r) :\equiv g(\boldsymbol{j})\boldsymbol{v}(r) z^{-J_0 J_1}.$$

Transition between random root functions, from $g(\boldsymbol{j})$ to $g(\boldsymbol{j}^{(2\pm)})$, is effected by the multiplier $z^{\pm J_0 J_1}$. Since both of these multipliers have identical spectral test valuations, we need to take only $z^{J_0 J_1}$. Now **Lemma 42** proves that transitions of elements of $g(\boldsymbol{j})\boldsymbol{v}(r)$ to those of $g(\boldsymbol{j}^{(2+)})\boldsymbol{v}(r)$ are given by $2r - 1$ generators

[37]Considerations may be extended to next-nearest N_2 or further neighbors at the formal level, though practicability horrifies us to proceed. Please see **Closing**.

$(I_2)\quad (d, z^{J_0 J_1}),\ (d, z^{J_0 J_1 \pm 1}),\ (d, z^{J_0 J_1 \pm 2}),\ \cdots,\ (d, z^{J_0 J_1 \pm (r-1)}).$

Proofs for other j_i's are obvious. **(End of proof of Theorem 43)**

8.6 Procedures of Non-Periodic Tuning

Generators in **Theorem 43** are tuned in various possible ways. Yet the order of tuning which starts over the stagger integer J_0 and proceeds sequentially over lattice size integers $J_1, J_2, \cdots, J_s$ will be the most adequate choice. This is because the procedures then

(1) start including only J_0, and then
(2) tune J_1 only without changing J_0, and
(3) proceed sequentially to $J_2, J_3, \cdots$ likewise.

We see actual aspects of procedures in the stated order.

Corollary 44. (Non-periodic tuning procedures) Start with the j_1 axis direction tuning, and proceed to $j_2, j_3, \cdots, j_s$ axes directions, all on inner lattice points.

(Tuning to the j_1-axis direction) First, take $2r-1$ generators:

$$(d, z^{J_0}),\quad (d, z^{J_0 \pm 1}),\quad (d, z^{J_0 \pm 2}),\quad \cdots,\quad (d, z^{J_0 \pm (r-1)}).$$

and perform their generalized 2nd tests in the noted order finding the appropriate increased or decreased J_0'.

(Tuning to the j_2-axis direction) We take $2r-1$ generators

$$(d, z^{J_0' J_1}),\quad (d, z^{J_0' J_1 \pm 1}),\quad (d, z^{J_0' J_1 \pm 2}),\quad \cdots,\quad (d, z^{J_0' J_1 \pm (r-1)}).$$

If some generator fails, we restart by replacing J_1 with $J_1' = J_1 + 1$ and resume tests again.

....

(Tuning to the j_s-axis direction) Generators to be examined are:

$$(d, z^{J_0' J_1' \cdots J_{s-2}' J_{s-1}}),\quad (d, z^{J_0' J_1' \cdots J_{s-2}' J_{s-1} \pm 1}),\quad (d, z^{J_0' J_1' \cdots J_{s-2}' J_{s-1} \pm 2)}$$
$$\cdots,\quad (d, z^{J_0' J_1' \cdots J_{s-2}' J_{s-1} \pm (r-1)}).$$

Tuning procedures are to end up with $J_0', J_1', \cdots, J_{s-1}'$ integers, with J_s unchanged in the present non-periodic boundary condition tuning.

(Proof) No inferences on the proof will be needed.

(End of the proof of Corollary 44)

The tuned forms of $J_0', J_1', \cdots, J_{s-1}'$ are not unique, and the lattice

$$G(\boldsymbol{j})' = J_1' \times J_2' \times \cdots \times J_{s-1}' \times J_s$$

have many possibilities in forms. Inventors feel that any choice of them will do well, but choices should be left to the wills of simulators. Some later considerations on time-dependent problems might be suggestive. Also, some later reference to the stagger integer J_0' might also be interesting from theoretical viewpoints.

8.7 Periodic Tuning for Random Initial Value Problems

Periodic boundary conditions are significant because of their able roles in giving various exact solutions, and as a means to entirely get rid of difficult problems arising from the boundary conditions. Let again take the lattice planned as $J_1 \times J_2 \times \cdots \times J_s$. Let $\boldsymbol{j}_{(i)}$ denote the starting boundary point of the lattice to the j_i-axis direction,

$$\boldsymbol{j}_{(i)} := (j_1, j_2, \cdots, j_{i-1}, j_i = 0, j_{i+1}, \cdots, j_s),$$

and let $\boldsymbol{j}^{(i)}$ be the end boundary point to the same j_i-axis direction,

$$\boldsymbol{j}^{(i)} := (j_1, j_2, \cdots, j_{i-1}, j_i = J_i - 1, j_{i+1}, \cdots, j_s).$$

The r-component random vector $\boldsymbol{v}(r) :\equiv (1, z, z^2, \cdots, z^{r-1}) \pmod d$ is as before. The random root functions $g(\boldsymbol{j})$ on boundaries are

$$g(\boldsymbol{j}_{(i)}) \equiv nz^{j_1 J_0 + j_2 J_0 J_1 + \cdots + j_{i-1} J_0 J_1 \cdots J_{i-2}} \times z^{j_{i+1} J_0 J_1 \cdots J_i + \cdots + j_s J_1 \cdots J_{s-1}} \pmod d,$$

$$g(\boldsymbol{j}^{(i)}) \equiv nz^{j_1 J_0 + j_2 J_0 J_1 + \cdots + j_{i-1} J_0 J_1 \cdots J_{i-2} + (J_i - 1) J_0 J_1 \cdots J_{i-1}} \times z^{j_{i+1} J_0 J_1 \cdots J_i + \cdots + j_s J_0 J_1 \cdots J_{s-1}} \pmod d.$$

This $g(\boldsymbol{j}^{(i)})$ has the expression

$$\begin{aligned} g(\boldsymbol{j}^{(i)}) &\equiv nz^{j_1 J_0 + j_2 J_0 J_1 + \cdots + j_{i-1} J_0 J_1 \cdots J_{i-2} + (J_i - 1) J_0 J_1 \cdots J_{i-1}} \\ &\quad \times z^{j_{i+1} J_0 J_1 \cdots J_i + \cdots + j_s J_0 J_1 \cdots J_{s-1}} \pmod d, \\ &\equiv g(\boldsymbol{j}_{(i)}) z^{(J_i - 1) J_0 J_1 \cdots J_{i-1}} \pmod d. \end{aligned}$$

We wish that the random vector function $g(\boldsymbol{j})\boldsymbol{v}(r)$ to be periodic across the boundary to directions of all spatial axes. What is necessary is that pairs of facing random numbers across the boundary are to be free of mutual correlation. Then necessary restrictions are that following $2r - 1$ *new* generators respectively have excellent generalized 2nd test valuation μ with $1 < \mu < 1.25$ for $1 \le i \le s$:

$$(d,\, z^{(J_i-1)J_0J_1\cdots J_{i-1}}),\quad (d,\, z^{(J_i-1)J_0J_1\cdots J_{i-1}\pm 1}),$$
$$(d,\, z^{(J_i-1)J_0J_1\cdots J_{i-1}\pm 2}),\ \cdots,\ (d,\, z^{(J_i-1)J_0J_1\cdots J_{i-1}\pm (r-1)}).$$

The following theorem summarizes one of our main attainments on random numbers distributed on lattices.

Theorem 45. (Conditions for random vector functions across boundaries to be correlation free) Take an MC generator (d, z, n) and let

$$G(\boldsymbol{j}) = \{(j_1, j_2, \cdots .j_s) |\ 0 \le j_i \le J_i - 1,\ \ 1 \le i \le s\}$$

be the spatial lattice $G(\boldsymbol{j}) = J_1 \times J_2 \times \cdots \times J_s$ with $J_i \ge 2$ for $1 \le i \le s$. Let

$$\begin{aligned} g(\boldsymbol{j}) :&\equiv nz\wedge\{J_0 k(\boldsymbol{j})\} \pmod d \\ &\equiv nz^{j_1J_0+j_2J_0J_1+j_3J_0J_1J_2+\cdots+j_sJ_0J_1J_2\cdots J_{s-1}} \pmod d, \end{aligned}$$

be the MC (d, z, n) random root function distributed on the lattice $G(\boldsymbol{j})$. As depicted in **Figure 14**, we denote $(\boldsymbol{I}_i)$ for the *relevant* MC generators that relate elements of the random vector function $g(\boldsymbol{j})\boldsymbol{v}(r)$ at an inner lattice point $\boldsymbol{j}$ to those on the nearest neighbor points to the j_i-axis direction. Let $(\boldsymbol{B}_i)$ finally denote *relevant* MC generators that relate elements of the random vector function on a boundary lattice point to elements of that on the $(\boldsymbol{N}_1)$ neighbor boundary point to the j_i- axis direction. Generators $\{(\boldsymbol{I}_i), (\boldsymbol{B}_i) |\ 1 \le i \le s\}$ are shown below:

(Proof) Proofs for generators of $\{(\boldsymbol{I}_i) |\ 1 \le i \le s\}$ are in **Theorem 43**. As generators in root functions connecting $g(\boldsymbol{j}_{(i)})$ of the starting boundary point and $g(\boldsymbol{j}^{(i)})$ at the end boundary point are related by $z^{(J_i-1)J_0J_1\cdots J_{i-1}}$, generators of $\{(\boldsymbol{B}_i)\}$ for random vector functions with $\boldsymbol{v}(r)$ follow from **Lemma 42**. ■

j_1-axis direction

$(\boldsymbol{I}_1)$ $(d, z^{J_0}),\ (d, z^{J_0\pm1}),\ (d, z^{J_0\pm2}), \cdots,\ (d, z^{J_0\pm(r-1)}).$

$(\boldsymbol{B}_1)$ $(d, z^{(J_1-1)J_0}),\ (d, z^{(J_1-1)J_0\pm1}),\ (d, z^{(J_1-1)J_0\pm2}),$
$\cdots, (d, z^{(J_1-1)J_0\pm(r-1)}).$

j_2-axis direction

$(\boldsymbol{I}_2)$ $(d, z^{J_0J_1}),\ (d, z^{J_0J_1\pm1}),\ (d, z^{J_0J_1\pm2}),$
$\cdots,\ (d, z^{J_0J_1\pm(r-1)}).$

$(\boldsymbol{B}_2)$ $(d, z^{(J_2-1)J_0J_1}),\ (d, z^{(J_2-1)J_0J_1\pm1}),\ (d, z^{(J_2-1)J_0J_1\pm2}),$
$\cdots,\ (d, z^{(J_2-1)J_0J_1\pm(r-1)}).$

j_3-axis direction

$(\boldsymbol{I}_3)$ $(d, z^{J_0J_1J_2}),\ (d, z^{J_0J_1J_2\pm1}),\ (d, z^{J_0J_1J_3\pm2}),$
$\cdots, (d, z^{J_0J_1J_2\pm(r-1)}).$

$(\boldsymbol{B}_3)$ $(d, z^{(J_3-1)J_0J_1J_2}),\ (d, z^{(J_3-1)J_0J_1J_2\pm1}),$
$(d, z^{(J_3-1)J_0J_1J_2\pm2}), \cdots, (d, z^{(J_3-1)J_0J_1J_2\pm(r-1)}).$

....

j_s-axis direction

$(\boldsymbol{I}_s)$ $(d, z^{J_0J_1J_2\cdots J_{s-1}}),\ (d, z^{J_0J_1J_2\cdots J_{s-1}\pm1}),$
$(d, z^{J_0J_1J_2\cdots J_{s-1}\pm2}), \cdots, (d, z^{J_0J_1J_2\cdots, J_{s-1}\pm(r-1)}).$

$(\boldsymbol{B}_s)$ $(d, z^{(J_s-1)J_0J_1J_2\cdots J_{s-1}}),\ (d, z^{(J_s-1)J_0J_1J_2\cdots J_{s-1}\pm1}),$
$(d, z^{(J_s-1)J_0J_1J_2\cdots J_{s-1}\pm2}),\ \cdots,$
$(d, z^{(J_s-1)J_0J_1J_2\cdots J_{s-1}\pm(r-1)}).$

Figure 14 Relevant generators for inner and boundary lattice points for periodic tuning

The remarkable new feature with the periodic tuning is that the lattice size integer J_i arises in this j_i-axis direction boundary condition. The ways of tuning on the **Figure 14** are again not unique, but now J_s is not free. We may suggest the simplest tuning process as follows.

Corollary 46. (Periodic tuning procedures) Random vector functions $g(\boldsymbol{j})\boldsymbol{v}(r)$ on the lattice $G(\boldsymbol{j})$ may be processed from **(1)** to **$(s+1)$** stated below.

(1) Tuning for J_0 in (I_1): Procedures for $(2r - 1)$ generators in (I_1) involves only the stagger integer J_0. These generators are the same as those that occurred in the non-periodic tuning, and we may proceed in the same manner looking for the adequate stagger integer J_0, either by increasing $J_0 \to J_0{}^* = J_0 + 1$ or by decreasing $J_0 \to J_0{}^* = J_0 - 1$. Repetitions of tuning will end with a $J_0{}^*$ that passes all of (I_1).

(2) Tuning of J_1 in (B_1) and (I_2): Next we consider the $2(2r-1)$ generators in (B_1) and (I_2), with the tuned $J_0{}^*$ and the new J_1 to be tuned. For both of relevant generators of (B_1) and (I_2) the tuning of J_1 is all routine; on failures we take increased $J_1{}^* = J_1 + 1$ and start anew 2nd tests until all tests are passed by a $J_1{}^* \geq J_1$.

....

(s) Tuning of J_{s-1} in (B_{s-1}) and (I_s): They give $2(2r - 1)$ generators with tuned $J_0{}^*, J_1{}^*, \cdots, J_{s-2}{}^*$. Increasing J_{s-1} to $J_{s-1}{}^* = J_{s-1} + 1$ on failures, we arrive at a $J_{s-1}{}^* \geq J_{s-1}$ that passes all of tests on these generators.

($s+1$) Tuning of J_s in (B_s): The remaining $2r - 1$ (B_s) generators are used to tune J_s, increasing it to $J_s{}^* = J_s + 1$ in failures.

(Proof) Descriptions of this **Corollary 46** itself are providing all of procedures and assertions. **(End of Corollary 46)**

8.8 Resumé for Ferrenberg-Landau-Wong Tests

We have used somewhat burdensome symbols $J_0{}', J_0{}^*, J_1{}', \cdots$ to discern tuned stagger or lattice-size integers. As we hereafter talk only on tuned data, we delete prime symbols on these integers.

What we discussed needed intricate descriptions to *build* tuned MC random number systems, that is, to find integers

r : number of random numbers to be on a lattice point,

s : dimension of the spatial lattice,

J_0 : stagger integer chopping MC sequence to J_0 pieces.

J_i : size integer of the lattice, $(1 \leq i \leq s)$.

Please also be careful the definition of the spatial lattice $G(\boldsymbol{j})$

$$G(\boldsymbol{j}) := \{(j_1, j_2, \cdots, j_s) |\, 0 \le j_i \le J_{i-1},\ 1 \le i \le s\}.$$

We disclose some of these tuned data systems here and in the next section, for the benefit of readers, to give here a short summary *how to use them*. Later, we shall discuss how to use *periodically tuned set of MC generators* for aims needing no periodic boundary conditions.

(a) We first fix the number r of MC random numbers with the independent disguise to be given on a lattice point. The dimension s of the successfully tuned lattice is limited to $s = 2$ at present, so that we only take this s.[38] We then distribute first r random numbers of the $k(\boldsymbol{j})$-th J_0-piece to the lattice point $\boldsymbol{j}$ by putting

$$nz\wedge\{J_0 k(\boldsymbol{j})\} \pmod{d}.$$

This finishes the setup of periodically tuned MC random numbers.

(b) If periodicity at lattice boundaries is not required, we may select any sublattice, typically in the $s = 2$ case,

$$G(\boldsymbol{j}) := \{(j_1, j_2) |\, 0 \le j_1 < J_1, 0 \le j_2 < J_2\}$$

of the periodically tuned $G(\boldsymbol{j})$. Note that we should retain J_1, J_2 in definitions by J_0, J_1, J_2 of the function $k(\boldsymbol{j}')$. This will serve for any boundary conditions, say free, fixed, reflexive, $\cdots$, keeping the distribution of MC random numbers as they were.

We, the authors, are not well trained in Dynamical Ising spin problems, not to mention Glauber or other dynamics. We somehow worked out the way to distribute r MC random numbers on spins with little nearest-neighbor correlations. Since disclosures of the next **Section 8.9** will make our MC random numbers to be readily applicable in tests of Ferrenberg et al. simply by replacing the random number routine, we hope those tests to be performed with tuned MC random numbers introduced here. Results will be a very stern test on #001 and/or #003, in line as Ferremberg et al. started

[38] If periodicity is not requested, successful tunings have been found also in cases $s = 3$.

their work. Is it really the problem of correlations of MC random numbers given to nearby neighbors? Or, the *hidden errors* reported have a different origin?

Now is the time to see the precise origin of errors, even if we may find all efforts of ours to tumble down.

8.9 Disclosure: Some Periodically Tuned Data

Guided by random initial value problems of Ferrenberg et al., we came up to see the possibility to distribute MC random numbers on the s-dimensional spatial lattice $J_1 \times J_2 \times \cdots \times J_s$ keeping correlation-free disguise of distributed r ($r \leq 3$ at present) MC random numbers between nearest neighboring lattice points. The basic strategy was to take every J_0-th MC random numbers

$$\{nz^{lJ_0} \pmod{d} \mid l = 0, 1, 2, \cdots, J_1 J_2 \cdots J_s - 1\}$$

as seeds or roots, sow them on respective lattice points and graft the length r ($r < J_0$) vector $\mathbf{v}(r)$ on roots. We expect clever choice of stagger integer J_0 and lattice size integers $J_1, J_2, \cdots, J_{s-1}$ or J_s by appropriate tests of extended 2nd-degree types. Tuning processes were, and will continue to be, laborious. Yet the results enable us to exploit them in a vast range of possibilities. The spatial lattice may have a void or voids with appropriately assigned boundary conditions. As we see in the next chapter, the results may also be extended to a distribution of MC generators on time-space lattices, which will also enable us to create new formulations of random walk problems. Expecting such applications, we present here periodically tuned 3 sets of J_0, J_1, J_2 with $(r, s) = (3, 2)$.

8.9.1 The MC Generator #003 on a Small Spatial Torus

We recall the setting. Consider a $s = 2$-dimensional torus $J_1 \times J_2$. Assume that its lattice point (j_1, j_2) is given $r = 3$ MC consecutive random numbers of #003 with good mutual independence. To the j_1-direction the lattice point has 2 nearest neighbors $j_1 \pm 1$. *As all lattice points* are distributed with MC random integers of the form $nz^k \pmod{d}$, (j_j, j_2) and its j_1-direction nearest neighbors are

j_1-axis direction

$(\boldsymbol{I}_1)$ (d, z^{J_0}), $(d, z^{J_0\pm1})$, $(d, z^{J_0\pm2})$.

$(\boldsymbol{B}_1)$ $(d, z^{(J_1-1)J_0})$, $(d, z^{(J_1-1)J_0\pm1})$, $(d, z^{(J_1-1)(J_0\pm2)})$.

j_2-axis direction

$(\boldsymbol{I}_2)$ $(d, z^{J_0J_1})$, $(d, z^{J_0J_1\pm1})$, $(d, z^{J_0J_1\pm2})$.

$(\boldsymbol{B}_2)$ $(d, z^{(J_2-1)J_0J_1})$, $(d, z^{(J_2-1)J_0J_1\pm1})$, $(d, z^{(J_2-1)J_0J_1\pm2})$.

Figure 15 Relevant generators connecting inner and boundary lattice points for $(r, s) = (3, 2)$ periodic tuning.

related by the MC generators (d, z^l) with some relevant power l. Such relevant generators were summarized by **Figure 15**. In the following, we reproduce **Figure 15** in the form relevant to the problem at our hand $(r, s) = (3, 2)$ classifying inner and boundary lattice points.

Tuning procedures of this system were performed by Naoya Nakazawa starting from J_0 in $(\boldsymbol{I}_1)$, then J_1 in $(\boldsymbol{B}_1)$, and so forth. Success was obtained as follows.

j_1-axis direction

$(\boldsymbol{I}_1)$ (d, z^{J_0}) : 1.21062052
$(\boldsymbol{I}_1)$ (d, z^{J_0+1}) : 1.10843200
$(\boldsymbol{I}_1)$ (d, z^{J_0-1}) : 1.04136399
$(\boldsymbol{I}_1)$ (d, z^{J_0+2}) : 1.24244511
$(\boldsymbol{I}_1)$ (d, z^{J_0-2}) : 1.11926379
$(\boldsymbol{B}_1)$ $(d, z^{(J_1-1)J_0})$: 1.11492279
$(\boldsymbol{B}_1)$ $(d, z^{(J_1-1)J_0+1})$: 1.16988254
$(\boldsymbol{B}_1)$ $(d, z^{(J_1-1)J_0-1})$: 1.18495608
$(\boldsymbol{B}_1)$ $(d, z^{(J_1-1)J_0+2})$: 1.04981696
$(\boldsymbol{B}_1)$ $(d, z^{(J_1-1)J_0-2})$: 1.13114224

j_2-axis direction

$(\boldsymbol{I}_2)$ $(d, z^{J_0J_1})$: 1.10142997
$(\boldsymbol{I}_2)$ $(d, z^{J_0J_1+1})$: 1.19744283
$(\boldsymbol{I}_2)$ $(d, z^{J_0J_1-1})$: 1.14919380
$(\boldsymbol{I}_2)$ $(d, z^{J_0J_1+2})$: 1.23600218

$(\boldsymbol{I}_2)$ $(d, z^{J_0J_1-2})$: 1.14824838
$(\boldsymbol{B}_2)$ $(d, z^{(J_2-1)J_0J_1})$: 1.15780383
$(\boldsymbol{B}_2)$ $(d, z^{(J_2-1)J_0J_1+1})$: 1.15679638
$(\boldsymbol{B}_2)$ $(d, z^{(J_2-1)J_0J_1-1})$: 1.07671814
$(\boldsymbol{B}_2)$ $(d, z^{(J_2-1)J_0J_1+2})$: 1.22991590
$(\boldsymbol{B}_2)$ $(d, z^{(J_2-1)J_0J_1-2})$: 1.16891715

If you only use random numbers of #003 with specifications

$$(r, s) = (3, 2), \quad J_0 = 125203390, \quad J_1 = 304, \quad J_2 = 1029,$$

then all of the 2nd-degree valuations noted above are realized with periodic boundary conditions.

Or, if you use #003 and the data (J_0, J_1, J_2) noted above, but on a smaller lattice $J_1{}' \times J_2{}'$ with, e.g.,

$$0 < J_1{}' \leq J_1, \quad 0 < J_2{}' \leq J_2,$$

your simulation works as on a non-periodically tuned spatial lattice $L' := J_1{}' \times J_2{}'$. You may also think of voids existing in L'.

8.9.2 Generator #001 Requires Different Toruses

Use of #001 MC generator necessitates a different set of integers $(J_0, J_1.J_2)$ for the torus $J_1 \times J_2$. A successful tuning is as follows.

$$(r, s) = (3, 2), \quad J_0 = 134215008, \quad J_1 = 1182, \quad J_2 = 287.$$

The 2nd-degree valuations are as follows.

j_1-axis direction

$(\boldsymbol{I}_1)$ (d, z^{J_0}) : 1.20916516
$(\boldsymbol{I}_1)$ (d, z^{J_0+1}) : 1.06512466
$(\boldsymbol{I}_1)$ (d, z^{J_0-1}) : 1.20697136
$(\boldsymbol{I}_1)$ (d, z^{J_0+2}) : 1.07149030
$(\boldsymbol{I}_1)$ (d, z^{J_0-2}) : 1.19109284
$(\boldsymbol{B}_1)$ $(d, z^{(J_1-1)J_0})$: 1.10735558
$(\boldsymbol{B}_1)$ $(d, z^{(J_1-1)J_0+1})$: 1.08727389
$(\boldsymbol{B}_1)$ $(d, z^{(J_1-1)J_0-1})$: 1.10554520
$(\boldsymbol{B}_1)$ $(d, z^{(J_1-1)J_0+2})$: 1.09778970
$(\boldsymbol{B}_1)$ $(d, z^{(J_1-1)J_0-2})$: 1.04201270

j_2-axis direction

$(\boldsymbol{I}_2)\ (d, z^{J_0J_1})$: 1.24828995
$(\boldsymbol{I}_2)\ (d, z^{J_0J_1+1})$: 1.22904438
$(\boldsymbol{I}_2)\ (d, z^{J_0J_1-1})$: 1.10817889
$(\boldsymbol{I}_2)\ (d, z^{J_0J_1+2})$: 1.07094361
$(\boldsymbol{I}_2)\ (d, z^{J_0J_1-2})$: 1.20262775
$(\boldsymbol{B}_2)\ (d, z^{(J_2-1)J_0J_1})$: 1.15507028
$(\boldsymbol{B}_2)\ (d, z^{(J_2-1)J_0J_1+1})$: 1.15972912
$(\boldsymbol{B}_2)\ (d, z^{(J_2-1)J_0J_1-1})$: 1.08417047
$(\boldsymbol{B}_2)\ (d, z^{(J_2-1)J_0J_1+2})$: 1.14741649
$(\boldsymbol{B}_2)\ (d, z^{(J_2-1)J_0J_1-2})$: 1.06550738

Random numbers of #001 used with specifications

$$(r, s) = (3, 2), \quad J_0 = 134215008, \quad J_1 = 1182, \quad J_2 = 287$$

gives a periodically tuned spatial torus. Also, these data used on a smaller lattice $J_1' \times J_2'$ with

$$0 < J_1' \le J_1, \quad 0 < J_2' \le J_2,$$

enable simulations as a non-periodically tuned spatial lattice, allowing for possibilities of voids. This tuning was also effected by Naoya Nakazawa.

8.9.3 Spatial Lattice Close to a Square with #001 Generator

This is also a torus built on the MC system #001 with

$$(r, s) = (3, 2), \quad J_0 = 1068627, \quad J_1 = 3975, \quad J_2 = 3986.$$

These data ensure with #001 the following generalized 2nd-degree valuations. Below we disclose generalized 2nd-degree valuations of all these generators.

j_1-axis direction

$(\boldsymbol{I}_1)\ (d, z^{J_0})$: 1.12053425
$(\boldsymbol{I}_1)\ (d, z^{J_0+1})$: 1.10131915
$(\boldsymbol{I}_1)\ (d, z^{J_0-1})$: 1.15877432
$(\boldsymbol{I}_1)\ (d, z^{J_0+2})$: 1.14547676
$(\boldsymbol{I}_1)\ (d, z^{J_0-2})$: 1.22402191

$(\boldsymbol{B}_1)$ $(d, z^{(J_1-1)J_0})$: 1.11692256
$(\boldsymbol{B}_1)$ $(d, z^{(J_1-1)J_0+1})$: 1.06406701
$(\boldsymbol{B}_1)$ $(d, z^{(J_1-1)J_0-1})$: 1.17604644
$(\boldsymbol{B}_1)$ $(d, z^{(J_1-1)J_0+2})$: 1.12920838 $(\boldsymbol{B}_1)$ $(d, z^{(J_1-1)J_0-2})$: 1.24327708

j_2-axis direction

$(\boldsymbol{I}_2)$ $(d, z^{J_0J_1})$: 1.24166987
$(\boldsymbol{I}_2)$ $(d, z^{J_0J_1+1})$: 1.09452210
$(\boldsymbol{I}_2)$ $(d, z^{J_0J_1-1})$: 1.05552764
$(\boldsymbol{I}_2)$ $(d, z^{J_0J_1+2})$: 1.17193249
$(\boldsymbol{I}_2)$ $(d, z^{J_0J_1-2})$: 1.12318220
$(\boldsymbol{B}_2)$ $(d, z^{(J_2-1)J_0J_1})$: 1.14352630
$(\boldsymbol{B}_2)$ $(d, z^{(J_2-1)J_0J_1+1})$: 1.10440618
$(\boldsymbol{B}_2)$ $(d, z^{(J_2-1)J_0J_1-1})$: 1.15533449
$(\boldsymbol{B}_2)$ $(d, z^{(J_2-1)J_0J_1+2})$: 1.08193080
$(\boldsymbol{B}_2)$ $(d, z^{(J_2-1)J_0J_1-2})$: 1.19249363

We repeat. The data of #001 with specifications

$$(r, s) = (3, 2), \quad J_0 = 1068627, \quad J_1 = 3975, \quad J_2 = 3986$$

may be used to realize periodic tuning on this spatial lattice. Also, the use of #001 with the data (J_0, J_1, J_2) on a smaller lattice $J_1' \times J_2'$ with

$$0 < J_1' \leq J_1, \quad 0 < J_2' \leq J_2,$$

will make your simulation to work well as a non-periodically tuned spatial lattice, allowing for possibilities of voids. This tuning was also accomplished by Naoya Nakazawa.

As to 3-dimensional lattice we yet have no successful periodic tuning. These details are too minute to be described here. If you would like to no other information on successful tuning, please freely and frankly ask HRF. We welcome your candid opinions, including criticisms.

Chapter 9

Random Number Fields on Time-Space Lattices

We have considered at length the problem of Ferrenberg et al., which stood on the viewpoint to *distribute MC random numbers on degrees of freedom residing on spatial lattice points.* Alternatively, we may consider problems that MC random numbers are defined as proper attributes of spatial lattice points, and physical particles or degrees of freedom perform random walks, staying accidentally on respective lattice points, colliding to one another in prescribed ways, and *feeling* environments set up by local random numbers. The system of random numbers in this picture will better be called random (vector) fields on time-space lattices.

We learned that procedures of ways of tuning, non-periodic or periodic, of **Chapter 8** may be realized to build correlation-free distribution of MC random numbers on spatial lattices. In the terminology of (r, s) tuning, little success was achieved for $(3, 2)$ type periodic tuning, while much more success was met in non-periodic tuning up to $(3, 3)$ types. The general (r, s) tunings were effected by taking the product of $\boldsymbol{v}(r) \equiv (1, z, \cdots, z^{r-1}) \pmod{d}$ to the root function $g(\boldsymbol{j})$ building:

$$g(\boldsymbol{j})\boldsymbol{v}(r) \equiv nz\wedge\{J_0k(\boldsymbol{j})\}(1, z, \cdots, z^{r-1}) \pmod{d},$$

Random Number Generator on Computers
Naoya Nakazawa and Hiroshi Nakazawa

ISBN 978-981-4968-49-2 (Hardcover), 978-1-003-41060-7 (eBook)
www.jennystanford.com

with the variable base function

$$k(\boldsymbol{j}) = k(j_1, j_2, \cdots, j_s) := j_1 + j_2 J_1 + \cdots + j_s J_1 J_2 \cdots J_{s-1}.$$

This system of tuned MC random numbers distributed on spatial lattice notably admits a natural extension to all time t, reserving the spatially tuned character, to what we call *MC random fields on time-space lattice*. We denote it as

$$\boldsymbol{f}(t, \boldsymbol{j}) :\equiv z^{rt} g(\boldsymbol{j}) \boldsymbol{v}(r) \pmod{d}, \quad t = 0, 1, 2, \cdots,$$

and exhibit neat properties noted below.

Theorem 48 (Time-Space Random Vector Field with r Components) Above introduced MC $\boldsymbol{f}(t, \boldsymbol{j})$ vector field with (r, s) type characterization has the following properties.

(A) If the MC (d, z) random numbers retain their 6 consecutive outputs with an excellent disguise of independence, random vector field $\boldsymbol{f}(t, \boldsymbol{j})$ with r components for $r \leq 3$ at time t reserve the excellent disguise of mutual independence to components of the field at the next time $t + 1$.

(B) If the initial MC random field $\boldsymbol{f}(0, \boldsymbol{j})$ is tuned to be correlation-free periodically or non-periodically between lattice points at spatial nearest-neighbors, *nearest neighbor non-correlation property* is ensured at any later time $t = 1, 2, 3, \cdots$ respecting the initial periodically or non-periodically tuned property at $t = 0$.

(C) If the original (d, z) MC random sequence has the usable period T, MC random field on the space lattice at time t never reproduces the same spatial random field in so far as $t < T/r$ holds true.

(Proof) **(A)** Obvious.

(B) At any spatial lattice point $\boldsymbol{j}$, the random vector field,

$$\boldsymbol{f}(t, \boldsymbol{j}) \equiv z^{rt} \boldsymbol{f}(0, \boldsymbol{j}) \pmod{d}$$

for any time point $t = 0, 1, 2, \cdots$, has one and the same time factor $z^{rt} \pmod{d}$. Hence the spatial non-correlation property of $\boldsymbol{f}(t, \boldsymbol{j})$ at any t is the same as the initial non-correlation property of $\boldsymbol{f}(0, \boldsymbol{j})$ which is set up by the initial tuning processes.

(C) At a fixed spatial lattice point $\boldsymbol{j}$ the above MC random number sequence $\{\boldsymbol{f}(t, \boldsymbol{j})| \, t = 0, 1, 2, \cdots\}$ on the time axis is just the original MC sequence folded at every length r. The random field $\boldsymbol{f}(t, \boldsymbol{j}) \equiv z^{rt} \boldsymbol{f}(0, \boldsymbol{j}) \pmod{d}$ on the spatial lattice point $\boldsymbol{j}$ for different time t are all different in so far as $t < T/r$ is true. ■

Inspired by the initial value problem of Ferrenberg et al., we concentrated on the problem of distributing MC random numbers on all degrees of freedom of a spatial lattice to realize correlation-free distribution on spatial lattice points located at nearest neighbors. We may likewise consider problems to distribute random numbers on particles *not* on every lattice point, typically on some degrees of freedom wandering in the spatial lattice, colliding with others by some prescribed law, or wandering in lattices of irregular forms or in lattices with voids. The traditional random walk problems proceed from the simple yet readily realizable binomial random numbers, and admirably proceeds straight to the Gaussian random number problems of diffusion processes. The MC random fields described in **Theorem 48** are in a sense *too good* compared to the binomial random numbers. Yet the feature may open up wider possibilities. The present authors are not qualified nor experienced to promote such broader applications. MC random numbers and MC random fields with their beautiful properties will need willing scientists, engineers and creators who will find new possibilities of simulations. We deeply wish to have contributions from them. We shall be honored if the present analyses could be of help to them.

Closing

We could select solid building blocks for MC random number generation out from our simple school mathematics with the powerful help of Sunzi's theorem and also of the tiny but truly beautiful flower given by Gauss. We would say that the figure of the technology of random number generation on computers is now wholly renewed. With confidence we close this monograph by presenting the following 2 propositions.

Proposition A Random numbers on computers should be exclusively generated by the multiplicative congruential (MC) way with the generator (d, z, n) constituted by a *composite* modulus d formed by *distinct odd primes* as $d = p_1 p_2$.

Proposition B The multiplier z should be selected by exhaustive tests based on regular simplex criterions.

We also stress the discovery of Naoya Nakazawa in **Figure 6** of **Section 7.1**, which enables us to obtain MC random numbers of real*8 precision with the decisive 10 times faster computing speed, *with all procedures within integer*8 algorithms*. This might well be said the true innovation effected by Sunzi's theorem in the random number generation technology on computers.

Everything that arose in our voyage was simply our primary school arithmetic of integers. Of course, we needed to learn about the contributions of great people, Euclid, Sunzi, Euler, Lagrange, Gauss, Galois and so forth anew, to become able to taste the abundant fruits noted above. We should also pay tribute to successful or unsuccessful efforts of many scientists of the 20th Century, and also to engineers whose works brought us to the present results through the prosperity of computers.

Once obtained, the knowledge of excellent random number generators is the common property of all of us human beings. Authors will be happy if people of this world could give their respective contributions to peaceful developments with logical reasoning and beliefs, with a firm grasp of structures of integer arithmetic, responding to the grace that has been presented to us by integers for so long times in the history of this universe. If this monograph could help people to appreciate the significance of the plain logical reasoning, the authors shall be all happy.

We reflect that the novel results to distribute random numbers, in particular MC random numbers, in spatial lattices as discussed in **Chapter 8**, or the possibility of distribution in the time-space lattices as discussed in **Chapter 9**, are neat and promising. Of course, we are not specialized in random walk problems, and we should wait for the research works of specialists or *creators*. We shall again be all happy if we could have criticisms. Though we believe that results of **Chapter 9** are neat and promising to discuss particles performing random walks with possible collisions, we should wait for contributions, or possibly criticisms, from ardent simulators and creators.

We may conceive of any s-dimensional spatial lattice. Its any 2 lattice points $\boldsymbol{j}$ and $\boldsymbol{j}'$ are connected by a lattice line on which an infinite number of lattice points are residing. In **Chapter 8** we crawled in mud, so to say, examining all relevant nearest neighbor generators by the brute force of linear algebra or performing exclusive tests. Yet these are all usual processes in any research, and should not be looked down, even though we sigh at the extreme beauty met in the arguments of Gauss on the unique prime decomposition of integers. Circumstances suggest that we should proceed to consider next and further nearest neighbors on the spatial lattice as well as nearest neighbors, and should construct non-correlated arrangements of MC random numbers, though the authors shiver at foreseeable difficulties, even though the approximation theorem of **Section 1.3** might well be waiting for our brave and clever fight against.

More frugally speaking, we should have MC random number generators which give 7 or more consecutive outputs passing tests. We eagerly wish that this objective will be taken over by people, and will be realized sooner or later, hopefully sooner, enriching our knowledge and technologies further.

Hirakata Ransu Factory (HRF)

Shakusonji-cho 28-18-103, Hirakata-shi, 573-0081, Osaka, Japan
hrf-hirakata@fmail.plala.or.jp

The Representative: Naoya Nakazawa, born in 1973

(2000) Entered the Master Course of Science, Osaka Prefectural University
(2005) Doctor of Science (Applied Number Theory) of Osaka Prefectural University
(2008) Jointly found *Multiplicative Congruential Approximation for Uniform Random Number Sequences*
(2024) Joint founder of the HRF, in charge of theoretical and numerical research

The Advisory Representative: Hiroshi Nakazawa: born in 1940

(1963) Entered the Master Course of Physics, Kyoto University
(1967) The Assistant to the Faculty of Science of Kyoto University, Physics
(1971) Doctor of Science (Physics) of Kyoto University
(1991) Takuma National College of Technology (TNCT), Professor of Applied Mathematics, began research works on random number theories and applications
(1995) Published the review textbook for TNCT, *The Mathematics of Uniform Random Numbers*
(2003) Retired from TNCT
(2008) Jointly found *Multiplicative Congruential Approximation for Uniform Random Number Sequences*
(2024) Joint founder of HRF in charge of theoretical research and general affairs

Index

Printed in the United States
by Baker & Taylor Publisher Services